Research, Invent, and Create Wealth

PETER JAMES KPOLOVIE

Copyright © 2024 by Peter James Kpolovie

Except as permitted under the U.S. Copyright Law, no part of this book may be reprinted by any means, graphic, electronic, or mechanical, including photocopying, recording, taping or by any information storage retrieval system without the written permission of the author, excluding the case of brief quotations embodied in critical articles and reviews with due referencing.

ISBN: 97-9832-480-892-1 (Paperback)
ISBN: 97-9832-481-041-2 (Hardcover)

USA: AMAZON KDP

Specific print information is available on the last page.

CONTENTS

Introduction v
1. Create Wealth With Research 1
2. Living In The Invention Age 17
3. Impact The World With Your Products 33
4. Be Renowned For One Thing – Making Inventions 45
5. Be Phenomenally Generous To Invent 55
6. Strive For And Attain Abundance By Inventing 65
7. Utmost Human Right Is The Making Of Inventions 73
8. Invest Extraordinarily To Invent 89
9. Extremely Crave Making Inventions 99
10. Research Sustains The World's Advancement 113
11. Practice Research Skills To Autopilot Level 129
12. Brain Boosting For Making Inventions 147
13. Change Is Golden For Making Inventions 169
14. Use Every Opportunity To Invent 187
15. Take Extreme Actions To Invent 199
16. Create The Future We Crave 211
17. Deviate From The Status Quo To Invent 225
18. Creation Of Value With Research 245
19. Gain The Greatest Freedom By Inventing 259

REVIEW REQUEST 271
REFERENCES 273

INTRODUCTION

The book is a clarion call for us to individually and collaboratively research, make inventions, create value, generate wealth, and advance the world. No person invents by doing things the same way others do. A person does not create, invent, or innovate something by doing it the exact way people expect of him.

Research is best defined as "making impossibility possible" (Peter James Kpolovie, 2023, 1). The answer to "What is Research?" is simply three words – "Making impossibility possible."

Research makes impossibility possible and solves unsolvable problems via invention, discovery, creation, or innovation. Only when a person empirically makes an invention, discovery, creation, or innovation that he has done research.

Without making an invention, discovery, creation, or innovation, a person has not done research. A person can only be said to have executed research when he made an invention, creation, discovery or innovation. The invention, discovery, creation, or innovation made must be of immense value to humankind.

People worldwide queue to electronically pay to benefit from or purchase the invented product, service, idea, knowledge, truth, or device.

The international marketplace determines the worth of an invention in terms of how much people gladly pay for the invention.

INTRODUCTION

A person who has yet to have a product that people readily pay for worldwide has not researched. Every renowned researcher in the world has made useful inventions, discoveries, creations, or innovations that automatically attract dollars.

Research is done for the creation of wealth and improvement of people's lot. Research is executed for the enrichment of the people and the inventor. The impoverishment of the investigator and the people is never the purpose of research.

This book thoroughly presents how the millions of preeminent researchers and prolific inventors on the planet think and act. Surely, every person who studies, practices, masters the book, and emulates how prolific inventors think and act will eventually execute research and make a very useful invention, discovery, creation, or innovation to better people's lives, improve society, and advance the world.

Research is the quest for new possibilities. It deals with the invention or discovery of replicable solutions to unsolvable problems.

Invention is the process or act of bringing into existence something that has never been made before. It is the bringing of ideas or objects together to create a novel device, process, or something useful.

Discovery is the act of finding something that has not been known before. It is finding a functional solution for the first time. It is finding what no one has known previously.

Each invention is equal to an impossibility made possible. When you invent, create, discover, or innovate something of tremendous value to people, research has been done. Something of great value to humankind gets invented with research. Something that truly meets the great needs of man, gets invented when you research.

Invention, creation, or discovery is evidence of impossibility made possible and unsolvable problem made solvable. The audience and users gladly pay to get and benefit from what you have done - the invention made and the value created.

Ask yourself these questions:
1. Do I want to be paid for the outcome of my research?
2. What if I receive automated payment for the results of my research?
3. What if the products of my research keep working for me, magnetizing dollars all through my lifetime?
4. Am I craving getting paid automatically for what I know?
5. Would I love to have some research products that people rush and pay for?
6. Would I like a major company to mass-produce the outcome of my research for the world?
7. What if the products of my research are massively commercialized on a large scale?
8. What if my research products inexhaustibly pervade and dominate the global marketplace?
9. Will I be self-fulfilled when I become renowned as the inventor who remarkably advanced the world?

Suppose the answers to these and similar questions are affirmative. In that case, you can achieve such cravings by researching or funding others to research and make valuable inventions.

How to research, invent, make impossibilities possible and turn unsolvable problems into solvable ones, and consequently attain wealth and prosperity are excellently presented in the book – **Research, Invent, and Create Wealth**. Get a copy of the book, study and apply it to achieve all you seriously aspire to.

We can create a better tomorrow. It is our duty, responsibility, and obligation to invent and create a blissful tomorrow for humankind.

INTRODUCTION

Have you ever visualized a better tomorrow for the majority of people on the planet? Have you ever wished to invent, discover, or create something to materialize the blissful tomorrow?

From today, embark on the journey to make the inventions a better tomorrow depends on. Execute research and invent the ideal future people crave. We must set out with our research and make the inventions upon which the tomorrow we crave depends.

We must pay the great debt we owe the future. Everything we could have and use today was invented, discovered, or created by people's research yesterday.

What have we individually created, invented, or discovered to guarantee a much better tomorrow for humankind?

The debt we owe, the greatest debt we owe, the only real debt we owe, is the creation of a far better future for all. We owe the future the making of inventions, discoveries, creations, and innovations that will bring about an incomparably better future than today. We must go all out with our research to fully pay that debt.

We must individually self-sacrifice for the good of all. Research demands complete selfless sacrifice and investment for the best good of humankind. Only a person who develops herself to the level of selfless investment in and sacrifice for the best good of all can execute research.

Research can only be done by a person who has developed herself for this purpose. Invest heavily in your self-development to execute or fund research, make inventions, create value, generate wealth, and advance the world.

A person must self-develop for research. Only a self-developed person for research can utterly seek:

1. Meeting man's needs.
2. Provision of solutions to unsolved problems.
3. Unraveling the truth and adding to knowledge.
4. Identification and resolution of unsolvable problems.
5. Discovery of novel cause-and-effect relationships.
6. Accomplishment of new possibilities.
7. Working in her alarm zone to make inventions.
8. Valuing the making of discoveries and inventions more than everything else, even her life.
9. Creating value with her inventions and achieving uncommon wealth, greatness, glory, and prosperity.
10. Advancement of the world by making the necessary inventions.

Peter James Kpolovie

CHAPTER 1

CREATE WEALTH WITH RESEARCH

A person creates wealth by inventing something of great value to the public, the more than eight billion people on the planet. The best and surest way a person can and should create wealth is by inventing, discovering, or creating value for humankind with his research.

1. Create abundant wealth with your inventions.
2. Invention-making rules the knowledge economy.
3. Be exceedingly inventive.
4. Create value and attain peak prosperity.
5. Pay the price, get the value.
6. Enrich others and become rich.

CREATE ABUNDANT WEALTH WITH YOUR INVENTIONS

The solution you create for an unsolvable problem is equivalent to the creation of wealth, limitless wealth, and abundant wealth for yourself and all the stakeholders in the invention. Every value a person's research creates for the world equals indescribable wealth he creates for himself and the other people involved in the commercial mass production and distribution of the product. Every person from across the world who benefits from the value you have created pays some dollars for it.

To attain and sustain great wealth on this planet, research and invent products, devices, ideas, services, solutions, or things of immense worth that people gladly rush and electronically pay some dollars for to get from everywhere and at every season and time globally. To create and sustain wealth, invent value that is greatly demanded.

Dear friend, would you want to attain wealth, greatness, and glory? If "Yes" is the answer, and it should be, all you have to do is invent or create highly valuable products, devices, ideas, services, and solutions. Each value you create with your research is equivalent to an endless stream of wealth, greatness and glory on to you.

There is an abundance of money in the world. The money in circulation only flows to the extremely few people who have created things of great value to humankind. The moment you create something of tremendous value, money flows to you and abounds you with wealth over time.

The quality services you render, the great ideas you create, the perfect solutions you invent, and the invaluable innovations you originate and accomplish for humankind are examples of value that magnetize wealth to you. Every new possibility you create for the betterment of humankind, improvement of society, and advancement of the world is a great value that attracts wealth to you and your associates or collaborators, as well as the beneficiaries. Such wealth keeps flowing to you ceaselessly and automatically pushes you to invest better in creating greater value.

All reasonable people everywhere in the world pay money to acquire things, solutions, and services that meet their needs and solve their problems. Please, be the superhuman who creates, invents, or discovers the things, solutions, and services that all others gladly and proudly pay for to meet their needs.

When you invent or create such values, you automatically become the centre to which money flows from everyone in every part of the world. Be the creator of the values people pay for. There is no better way to amass wealth than the creation of value.

People can only get the value, product, solution, idea, and service they pay for. When you become the creator, inventor, or discoverer of a solution, product, value, device, idea, or service, the payments shall be made to you. This accounts for why the best investment that a person could make is in the invention and creation of great value for humankind. Therefore, it is perfect to invest heavily in making inventions of amazing value to humanity.

Anyone who truly needs wealth gets it, perhaps even in abundance, by making substantially valuable products, goods, ideas, inventions, and solutions and rendering incredibly useful services to humankind. The universe rewards such persons with boundless wealth.

Exceptionally crave the making of extremely useful inventions with your research. Exert utmost action at all costs to get the inventions excellently made. Then, nature, God, and the universe shall unfailingly reward you with unimaginable glory, greatness and wealth.

There has never existed on the planet any wealthy man who attained it without first inventing or creating great value for humankind. Each of the wealthy persons outlined below (Fobes, 2024; Nichesss, 2024; JUSTIA Patents, 2024; United States Patent and Trademark Office, 2024) became wealthy by creating enormous and substantial value for humankind. They each created things or services of exceptional worth for which people from everywhere on Earth paid to benefit from the created value.

1. Elon Musk (Tesla, SpaceX).
2. Steve Jobs (Apple, iMac, iPad, iPhone).
3. Bill Gates (Microsoft).
4. Benjamin Franklin (Polymath, lightning rod, bifocals, and Franklin stove).
5. Henry Ford (Eponymous carmaker, Ford Motors Company).
6. Samuel Colt (Revolver for mass-production).
7. James Dyson (Cyclonic Separation – Dirt and vacuum cleaner).
8. Gary Michelson (Spine and other Surgical equipment).
9. Richard Arkwright (Industrial Revolution).
10. Alfred Nobel (Dynamite inventor, Nobel Prize in literature).
11. Thomas Alva Edison (General Electric, plus over 1093 US patents).
12. Jeff Bezos (Amazon).
13. Bernard Arnault (LVMH – Moet Hennessy Louis Vuitton for luxury goods).
14. Mark Zuckerberg (Facebook).
15. Steve Ballmer (Microsoft).
16. Warren Buffett (Berkshire Hathaway).
17. Larry Ellison (Oracle).
18. Larry Page (Google).
19. Sergey Brin (Google).
20. Mukesh Ambani (Diversification).
21. Michael Dell (Dell Technologies).
22. Michael Bloomberg (Bloomberg LP).
23. Jensen Huang (Semiconductors).
24. Gautam Adani (Infrastructure, commodities).
25. Jim Walton (Walmart).
26. Rob Walton (Walmart).

27. Alice Walton (Walmart).
28. Zhong Shanshan (Beverages, pharmaceuticals).
29. Giovanni Ferrero (Nutella, chocolates).
30. Zhang Yiming (TikTok).
31. Colin Huang (E-commerce).
32. Shunpei Yamazaki (Semiconductor Energy Laboratory, holds over 11,833 U.S. patents).

Suppose a person unreservedly takes all the risks, persistently makes all the investments, relentlessly makes every required sacrifice, and committedly exerts all necessary excessive actions nonstop to make a crucial invention. In that case, he triumphantly overcomes all the unbeatable odds that have prevented everybody else from making the target invention.

Next, nature unveils the person the perfect solution to the problem that has all along been concluded to be unsolvable. Then, illimitable wealth ceaselessly flows to him from all directions as an additional reward for his meticulous research and the accomplished crucial invention.

A person only invents a thing after he has taken every associated risk beyond all human imagination, invested everything (no matter what) in it, surmounted all the insuperable, and persistently taken all the necessary extreme actions for excellently making the invention. When the target product gets superbly invented, it gets ineliminably demanded.

People making the demands pay some dollars to get the product. Such payments from across the world inexhaustibly pour in. Consequently, the inventor, the collaborating manufacturers and the effective distributors get massively enriched. That is why each essential invention achieved enriches some people, companies and multinational corporations, improves individuals and society and advances the world.

Inventions are made for the enrichment of the stakeholders. Inventions are never made to impoverish the inventor, investors and the other stakeholders. Inventions only better people, improve society and advance the world. The opposite is never the purpose of any invention-making journey.

Embark on making an invention for your wealth creation, enrichment of the collaborative manufacturers and distributors, and fantastic improvement of people's lot. Do everything it takes to invent or create value for humankind. In return, amass wealth for yourself and the limitless number of people participating in the product's mass production, distribution, or marketing. Ultimately, the invention enriches and improves the end-users as it overwhelmingly meets their needs.

INVENTION-MAKING RULES THE KNOWLEDGE ECONOMY

In the knowledge economy we live in, research upon which the creation, invention, discovery, and innovation of knowledge depend is the ultimate. Therefore, creating value and inventing extremely useful ideas, knowledge, products, solutions, devices, technologies, and services should be all you crave. Plus, do everything to incredibly achieve your craving. Such achievement shall earn you an automated stream of endless wealth, glory and greatness.

With or without a college degree, employment, or a company, everyone who truly wants can research, invent, discover, create, or innovate a solution or something of great worth and transform it into an ever-flowing wealth. Committedly embark on such invention-making journey and excellently achieve the target invention.

All a person needs to get started is to acquire a copy of this book (which you have already done), study it (you are marvellously doing that), and apply the guiding information therein. When fully done, the results shall be the making of a highly valuable invention, discovery, or creation of a product, service, knowledge, or solution that people worldwide will gladly rush and pay some dollars to benefit from.

With this book, anyone can execute research and turn the products into an unending streaming wealth. With the committed application of the book, a person can make useful inventions, creations, discoveries, or innovations and get paid globally for his created value.

Never has it been as easy as in our Invention Age and digital world for a person to research, make a useful invention, turn it into a digital product, and magnetize ever-streaming wealth annually, monthly, weekly, or daily. The creation of value rules the flow of wealth in the knowledge economy. Invent what people crucially need from across the world, and wealth shall accordingly flow to you.

We are living in the best time, the Invention Age, a digital knowledge age, a digital economy age, an e-commerce age, and an information and communication technology age. Thanks immensely to all the people whose relentless research works have made the inventions that have given us the most suitable era for easy value creation and wealth generation.

We only individually need to continue advancing the world by making necessary inventions for a much better tomorrow. We must not fail to substantially contribute our quota by making inventions for the world's advancement with our research today.

In this Invention Age with information and communication technology, we can most easily turn everything, just anything we want, into a source of wealth with our research. Never a thing that we dedicatedly embark on inventing which we cannot have marvellously invented when we do all the needful. We never invent anything with our research that cannot be transformed into an amazing wealth generator. Every invention a person makes can be transformed into an astonishing wealth generator in this Invention Age.

We only need to meticulously execute the necessary research, digitalize the indelibly valuable product, and disseminate it electronically. Then, the ceaseless inflow of dollars shall keep coming to us.

Every thriving seven-figure, nine-figure, or ten-figure business annually in the world started with research, developed and progressed with research, and is sustained with research. Please begin your own. Invest heavily in research today for the invention of the highly valuable products upon which your seven-figure, nine-figure, and ten-figure business annually in the future depends.

Make crucially useful inventions with your research today. Distinctly stand out worldwide with your highly valuable and ubiquitously demanded invented products, solutions, and services. Then, over time, you shall progress to and remain in the top one per cent of inventors who dominate the global marketplace with their inventions.

With your inventions, attain and live a superb life. Help others with your inventions to accelerate to their maximum potential and attain superb lives. Mentor countless others with easily replicable examples to committedly pursue and successfully create new possibilities with their research by inventing products, services, and solutions that have long been concluded to be impossible in the world.

BE EXCEEDINGLY INVENTIVE

Be exceedingly inventive. Strive desperately at and attain an abundance of inventions with your research. An uninventive life is not worth living.

Like every renowned and prolific inventor on planet Earth, you have the power, potential, and special divine spark to most successfully research and invent whatever you want. Enthusiastically and committedly embark on that unique mission, responsibility, duty, and obligation of meticulously making an extremely valuable invention to better human lives, ameliorate society and advance the world.

Accept total responsibility for your life and your invention-making journeys to ensure excellent completion every single time. Fabulously accomplish each research work you embark on. Get the target invention made each time you embark on research.

Your inventions, products, solutions, and services should make a positive difference in the world. Fantastically making the invention your research aims at should be much more valuable to you than any amount of money and everything else. Get the target invention excellently made at all costs in each invention-making journey you embark on.

Success, prosperity or wealth is having what a person really wants in life. Money cannot buy the most important thing we want and should want in life. That thing is making indispensable and highly valuable inventions. Every person who really wants can successfully achieve a valuable invention with his research.

The awesome making of an indispensably useful invention is achievable if a person wholeheartedly chooses it and works relentlessly at attaining it. But suppose somebody does not choose it or does not believe it and refuses to work on attaining it. In that case, he will definitely fail in making any invention.

A person who chooses to make an invention must persistently

work on developing himself, his attitude, habits, beliefs, and action exertion. He has got to do everything to move ahead of everybody else on Earth to be the first to make the target invention.

Everyone faces seemingly indomitable obstacles in the invention-making journey. But do you really want to accomplish the invention? Yes. In that case, you have got to keep moving ahead and overcoming every one of the obstacles, no matter what, by doing every damn thing that the making of the invention demands until it is extraordinarily accomplished.

With untold passion, do everything making the invention demands. Love making the invention so much that you could invest everything else in it and make all the sacrifices that making the invention demands. Then, the invention gets marvellously made. When excellently made and disseminated, the invention turns into an overflowing stream of never-ending wealth for you and all the other concerned people.

CREATE VALUE AND ATTAIN PEAK PROSPERITY

Execute research that creates or invents something of boundless value. Put a price on the value. People worldwide who need the created value will pay the price and get the value.

Make extremely valuable discoveries or innovations with your research. Place a price on each of the products. People will gladly pay the price willingly and get the value.

The greatest and most effective principle for the attainment of peak prosperity is this:
- First, create value.
- Second, put a price on it.
- Third, receive payment people make to get the value.

The invention of something of monumental value alone gives a person unimaginable fulfilment. The self-fulfillment and self-satisfaction alone are prosperity attained.

The price the international marketplace assigns to the value you created alone raises your self-worth and your worthiness to the peak. This is a priceless prosperity you attain with your invention or discovery.

When people from everywhere on Earth pay to get the value you created, you again ascend to the climax of prosperity. Prosperity and wealth depend on the volume and worth of value a person creates with his research.

To be truly prosperous, great and glorious, you have got to create value. You can best create value by inventing, discovering, or innovating something of immense usefulness to people worldwide with your research.

The created value satisfies your cravings and gives you an uncommon sense of living a prosperous life. Crave the making of an invention. Invest everything it takes to get the invention made. Then, have your craving fulfilled by inventing the target product.

With the attachment of a price to your created value by the global marketplace, you experience another level of prosperity. It is a level of prosperity that nothing else can give.

The payment people universally make to benefit from your created value crowns your prosperity with timeless rhodium. When people ubiquitously pay to gain from the solution, product, or service you invented, your prosperity, glory, greatness, and wealth reach the peak that only inventors could attain.

Don't you want to be extremely prosperous?

What are you still waiting for if you truly want to be so prosperous?

Embark on the research that will create, invent, or innovate something of astronomical value to humankind. The value you create or invent will automatically make you experience boundless wealth and extreme prosperity.

The only thing that gives a person, family, collaborative team, company, organization, multinational corporation, or nation boundless wealth and prosperity is the inventions they make and the value they create. To be prosperous, invent things of enormous usefulness. To attain and maintain abundant wealth and prosperity, create value upon value.

The prayer or wish "Long life and prosperity" or "Be prosperous" when people clink their glasses does not, cannot, and will never make any person, organization, company, or state prosperous. The only thing that gives prosperity is the value you create. Only the value a company, organization, multinational corporation, or nation creates gives her endless prosperity and great wealth.

PAY THE PRICE, GET THE VALUE

Pay the price. Then, get the value. To get value, you must pay the full price for it. Never has any person created value without first paying the price for it. Similarly, never shall a person create any value without first paying the full price for the value creation.

A person must pay the price, usually a huge price, before he can invent or create something of great value for humankind. Without unreservedly investing in and thoroughly executing the research that making a particular invention depends on, a person cannot create the target value.

Invest everything in and about you in rigorously executing the research upon which the value creation depends. Only then, can you create the type of value so essential that people from across the globe gladly pay some dollars to benefit from.

Pay the price with your research to create the essential value from which countless people will willingly let go of some hard-earned dollars to benefit.

What you pay for is what you get. In life, a natural rule is that a person gets only what he has paid for.

Do you want to create great value with your research? Yes. Pay the full price for getting the research that creates the great value done.

Please, squarely take the challenge of getting the research that creates value executed. From today, start investing time, resources, treasure, and talent in thoroughly executing research to create a universally useful product, service, or solution.

Pay the huge invention-making price. Get the value created. Next, people from everywhere in the world will individually pay a small amount of dollars to gain from the value your research has created.

For a crucial value created, people endlessly pay to benefit from it. Over time, the inventor gets a million times more than all he invested in creating the value.

The earnings of such payments could be referred to as residual incomes. Residual income is when the value you create now with your research continues to endlessly magnetize the inflow of dollars to you in the future, even when you are not making other inventions. It is when you make an invention once, and the payment people worldwide make to you to benefit from the invented product keeps coming over and over endlessly.

The essential rule remains:
1. Pay the price and get the value.
2. Pay the price for making an invention with your research.
3. People across the globe shall endlessly pay the price to benefit from the value you have created.

ENRICH OTHERS AND BECOME RICH

Prosper other people and become prosperous. Enrich others because your becoming wealthy depends on it. Never has a person become prosperous without first prospering others. Every truly wealthy person first made other people get rich.

You can best prosper others with the value you create. The best way for you to enrich others is through the inventions of irresistible value you make with your research.

When you enrich and prosper others with your created value, prosperity, wealth, greatness, and glory flow to you like an endless, unstoppable stream. You can only attain abundant wealth and prosperity after creating irresistible, indispensable value that satisfies the needs of humankind.

Invest your resources unreservedly in research to get something of great value invented. Then, you will get paid for the value you created. This way, individuals, society, the economy, nation and the world advance because of the value you have created. And your created value keeps magnetizing endless wealth to you.

The value we create with our research could be services, ideas, products, solutions, devices, technologies, or knowledge that were not considered possible before our research. Create such value. And wealth with true prosperity shall get attracted endlessly to you in abundance.

We attain true wealth, greatness and glory only with the value we create. By adding to the prosperity of other individuals, societies, and countries with the value we create, greater prosperity automatically flows to us.

The rule for the achievement of true success, prosperity and uncommon wealth is simply this:
1. Research and create value to enrich and prosper others, society, humankind, and the world.

2. In return, you automatically achieve overwhelming wealth, prosperity, success, self-fulfilment, and life mission accomplishment.

To attain peak potential, a person must add value to other people, society and the world. It is the value we individually add to society and the world through our research that gives us the greatest sense or feeling of self-fulfilment, life mission accomplishment, wealth, glory, and greatness.

A person can only achieve all-time greatness with the value his research created. Make an invention of immense worth. Then, you attain true greatness, glory, wealth, and prosperity in return.

CHAPTER 2

LIVING IN THE INVENTION AGE

We are living in the Invention Age. The Invention Age is the Age of invention making. It is the Age of value creation. The Invention Age is that which virtually every advancement in the world depends on the inventions we make. Nothing in human history has ever moved as fast as and had a great positive impact on the world's advancement as the making of inventions.

1. We are living in the Invention Age.
2. Wealth in the Invention Age.
3. Demonstrate God's qualities.
4. Become a classic inventor.

WE ARE LIVING IN THE INVENTION AGE

The making of inventions alone advances the world. Invention making has become so central that every improvement in the world is occasioned by the inventions we make.

The Invention Age is the era in which humankind's thoughts are predominated with the full realization that only the making of inventions creates value and advances the world. In the Invention Age, man completely realizes that the world's future is determined by the inventions we make today.

The Invention Age is when humans seriously accord the

making of inventions its rightful place in the scheme of events. The Invention Age is the era in which every individual not only craves but actually embarks committedly on creating value by making crucial inventions.

The solution to each of man's problems could best be found in making more inventions. Therefore, the Invention Age is predominated by the new consciousness that humankind can solve each of its persistent problems by making many more inventions.

Every problem of humankind that has been solved was achieved mainly by making inventions that dealt with the problem and made it a thing of the past. With all certainty, the problems man still faces can each be solved once the right inventions are made for the purpose.

In the Invention Age we live, invention-making has so revolutionized the world that only one or few yet-to-be-made inventions separate man from living a life free from each problem. In other words, we only need to make some right inventions to get liberated from each problem.

For instance, the invention of Information and Communication Technology (ICT) alone has solved a great deal of man's ICT-related problems. Information and Communication Technology has seen the making of countless priceless inventions for the advancement of the world. Information and Communication Technology is just one of the inventions in the Invention Age. ICT alone has given rise to innumerable inventions that are further revolutionizing the world for the best.

We just cannot do without the continuous making of inventions to better meet the needs of humankind. Suppose we truly desire deeply to live much better lives. In that case, we must research tirelessly today and make the inventions upon which a better life tomorrow depends.

Our future depends upon the inventions we make today. If we need a blissful future, we must make all the necessary inventions today to actualize that future we crave.

Currently, we must individually think invention-making thoughts. Committedly exert all the actions for making the inventions that will unfailingly create the type of future life we very strongly desire.

Whatever invention idea you think about the most, focus on making the most, and exert the greatest actions than everyone else on making it, will definitely get invented. Get started and work on it until the amazing completion of the incredibly valuable invention for the betterment of people, improvement of society, and advancement of the world.

Now is the auspicious time to begin and continuously work unstoppably on making the invention we crave the most until it is wonderfully actualized. Never think that your time to make an invention has passed.

It is still possible to self-improve and make any invention you crave the most at each time. It will be too late if we do not start now to self-improve for making inventions. Right this moment is the auspicious time for self-development to make crucial inventions.

Acquiring this book and reading up to this point is evidence that you are seriously on course. Congratulations. Keep it up, please.

Now, in the Information Age, is the best time for a person to learn and self-develop. Learn new ways of thinking. It is the right time for each person to begin and continue accelerating actions for making the inventions of their cravings.

Complete one invention and embark on another to a successful completion. Keep the invention-making process on, one after another, till you become a prolific inventor.

The opportunities to invest in self-development to change our course, set our sights on new horizons, and make many more inventions never end. Now is the right moment to examine where we are regarding existing inventions, where we ought to be, and where we should go on the invention-making continuum for greater success, wealth, health, happiness, and prosperity.

Now is the auspicious moment to embark on and work ceaselessly on the invention-making journey until an awestruck completion. Please embark on the journey of making an essential invention until the most successful completion.

Unlock your creative and invention-making potential with your research works. Then, achieve true prosperity by making the target inventions, discoveries, innovations, and creations each time.

The greatest national and global security risk in this Invention Age is the failure to make inventions and discoveries of great worth with our research. Nothing, absolutely nothing else threatens the world's advancement than the refusal or failure to make inventions, creations, and innovations with our research. Doing research that does not invent a thing of great worth is the biggest threat to humankind, global peace, and advancement.

Kpolovie's books are written to put an end to all such threats by helping people to execute research that makes crucial inventions, discoveries, creations, and innovations to best meet humankind's pressing needs and advance the world.

The books have helped establish four hours reading, writing, and experimenting as daily routines, activities that must be done every day. The books have created cravings for making inventions with our research. The cravings have created strong cues and rewards mindset in the users.

Simply: Execute research. Research makes inventions. Inventions get rewarded with the desired wealth.

People across the globe have developed an invention-making

mindset. They are researching and inventing much more than ever before.

Please research, invent and get paid for the inventions. Launch simple advertising campaigns that trigger irrepressible cravings for the products. Let the campaigns convince the audience and consumers to acquire and use your products every day.

Gaining your freedom from the status quo and making a crucial invention is the timeless and priceless wealth that a person should seek and attain via his research. The satisfaction of the craving and determination to make a more valuable invention depend on one's self. Living a prosperous life of abundance can only be achieved through meticulous research execution, and making highly valuable inventions.

Please live up to the demands of the Invention Age. Make essential inventions. Create indispensable value. Plus, attain the abundant prosperity it brings.

WEALTH IN THE INVENTION AGE

Emphatically, we live in the Invention Age, the Age of making inventions. The Invention Age in which we live is the era that what matters the most is the making of inventions. It is the Age that inventions made determine everything else. Wealth in the Invention Age is measured by inventions made and values created.

Having more than one mass-produced and globally commercialized invention is equivalent to having and living an abundant life. For centuries, America has been conclusively considered a land full of opportunities (Alamieyeseigha and Kpolovie, 2013). This assertion is much truer in this Invention Age than ever before.

In this Invention Age, there is an abundance of opportunities everywhere, and in everything, at every time, for every person in the United States, for making inventions and creating value. Many more people than ever have keyed in, are keying in, and will continue to key into using the abundant opportunities for value creation.

Very long ago, before the Invention Age, ownership of large amounts of land was a major and greatest source of wealth. But later, when the making of inventions took centre stage, everything changed for the better.

From the Invention Age forthwith, the greatest source of wealth in America, other developed nations, and even in some Third World countries is the making of valuable inventions. Never again shall there be a millionaire or billionaire in the world who does not own some inventions.

It is the era of inventions, the Invention Age that what counts the most is the making of inventions. In the Age of Inventions, to be globally renowned as a millionaire or billionaire, one must own some inventions, creations, innovations, or discoveries. Wealth in the Invention Age is and will continue to be measured primarily by the volume and value of inventions accomplished.

Sustainable wealth is a function of inventions made. Tremendous riches in the Invention Age can best be produced or amassed by the value, worth and volume of inventions made. Countries with most inventions are rich, while countries with few inventions are poor.

Abundant wealth, power and vast financial holdings depend and shall continue to depend primarily on the inventions made and owned. Relationships and business alliances shall be formed not based on raw materials owned but on the inventions, discoveries, creations, and innovations made.

People, states or nations, and continents that merely own land and raw materials shall remain poor. Some Third World countries are good examples of such poverty-inflicted countries.

They easily afflicted themselves with poverty by their refusal to prioritize value creation and invention making.

In contrast, the people, states or nations that dwell on making inventions, discoveries, creations, and innovations shall flourish in abundant wealth. The United States of America, France, China, Germany, United Kingdom, India, Canada, Japan, Mexico, Spain, Switzerland, Italy, Russia, Australia, Sweden, Netherlands, South Korea, Finland, Singapore, Ireland, Brazil, United Arab Emirates, Israel, Luxembourg, and Denmark are some leading inventive, creative, and innovative nations renowned in the Invention Age for their invention-based wealth.

We have today to decide, to choose between living in abject poverty and having wealth in abundance. If living in abundant wealth is what we have chosen, we must invest heavily and work wholeheartedly to make inventions.

With research alone, inventions are made. Therefore, we must committedly invest everything in and about us in execution of research for the making of all the needful inventions.

Choose invention-making and attain wealth in abundance. Conversely, refuse or neglect the making of inventions today and live a life of abject poverty tomorrow.

With the inventions we make today via our research, the type of future we crave gets closer to us. Research exceedingly today and make all the valuable inventions, discoveries, innovations, and creations that a blissful tomorrow depends on.

In this Invention Age, virtually everyone has opportunities to research, invent and create abundant wealth. Capitalize on it. Research meticulously and make tremendously useful inventions for humankind. Mass-produce the products, services, solutions, ideas, devices, technologies, or knowledge. Commercialize the inventions for all those in need to easily acquire them electronically with a touch of a few buttons on the computer.

There is, indeed, no better way of creating wealth in this Invention Age than making valuable inventions to improve people, better society and advance the world. Please invent things of great value and positively impact the world. Let your products help countless people accelerate faster toward attaining their maximum potential.

DEMONSTRATE GOD'S QUALITIES

The Invention Age requires us to create value as a demonstration of God's creative nature in us. We are individually created in the image and likeness of God to continue with God's creative work on Earth via our research.

God is the creator and sustainer of the universe. We have got to create things of great value. The value we create with our research is the expression of the extent to which we continue God's creative work on Earth. With the value we create with our research alone, we demonstrate the magnitude to which we sustain the world.

The sustainability and advancement of the world depend on inventions we make for the purpose. Each person is expected to demonstrate God's unique qualities and characteristics as evidence of fulfilling the purpose for which he was created in God's image and likeness.

In the Invention Age, we fully realize that we must utilize God's creative spark in us by being highly creative, inventive, and innovative with our research. For this reason, we do everything it takes to maximally utilize our infinite potential for solving all problems and making possible, humanly impossible things with our research.

God is omniscient. He is all-knowing. Please read four hours daily, write four hours daily, and experiment four hours each day to become all-knowing as much as possible. Do everything that will make you learn more and know much more. The greater the knowledge mastery you acquire and exhibit in any profession, the more you become like an omniscient God.

God is immortal. He lives forever. God never dies. He cannot be destroyed. Please research and make indelibly valuable inventions for humankind. A person's undestroyable and indelible creations, inventions, innovations, and discoveries immortalize, idolize and heroize him on Earth. Create permanent value, perpetual value, with your research.

God is omnipotent. God is all-powerful. Everyone has potential, talents and unique gifts that, if fully utilized, will make him gravitate towards all-powerfulness. Dedicatedly and committedly execute research with all your potential, talents and unique gifts. When you do so, the products of your research will exhibit the magnitude to which you are omnipotent. Wonderfully good things that all other human beings cannot make possible, you can make possible with your research.

A problem that all others have yet to be able to solve can get a perfect solution provided for with your research. You have demonstrated God's omnipotent character when you create new possibilities (turn things that, before your investigation, have been concluded by all human imaginations to be impossibilities) to become possibilities with your research. With God's omnipotent spark in you, you can create, invent or discover any valuable thing with your research.

God is omnipresent. The Almighty is present everywhere, always at the same time. Let the inventions, innovations, creations, and discoveries your research make be so extremely valuable that they get ubiquitously disseminated all over the universe. When a person's inventions or products are so important that they pervade the whole world at all times, the person has demonstrated the extent of his omnipresence.

Please thoroughly research. Invest everything in and about you in the making of excellent inventions. Make exceptionally useful inventions that get diffused throughout the world and remain everywhere at all times and simultaneously in the universe. Then, you have demonstrated the magnitude of your omnipresence.

When your products constantly dominate everywhere at all times, you and your inventions get known by countless people across the world. Then, you accomplish omnipresence. Let your inventions constitute a brand that is inexhaustibly available and demanded round the clock everywhere globally.

Exert extreme actions in your research execution and make products that accomplish your golden and godly mission of being everywhere simultaneously at all times. Create exceptionally valuable products, inventions, ideas, services, devices, discoveries, solutions, and brands that are universally ineliminable, illimitable, inexhaustible, and ubiquitous. When you accomplish this godly feat, you have exhibited that you were, indeed, created in the image and likeness of God. Kindly do everything, no matter what, to attain this great feat as the Invention Age demands.

With your research, provide irrefutable solutions to persistent problems of the masses worldwide. Let the solutions created by your research be most appealing, satisfying, unfailing, and easy to acquire and apply by the masses universally. When your invented solutions are so impactful, everyone who needs them can purchase them electronically, use and get the problem solved.

The products of your research should most successfully meet the needs of everyone who cares throughout the universe. This way, the international marketplace easily gets dominated by your products. The more people all over the world who acquire, correctly apply and solve their problems with your products, the better.

The research a person embarks on should aim at creating a solution to a currently unsolvable problem that persistently faces the masses across the globe. On outstanding attainment of the research goal, people ready and willing to acquire and apply the invented solution, should be in every part of the Global Village.

The created solution should be mass-produced and disseminated in collaboration with multinational corporations everywhere on the planet for all those in need to easily acquire, apply correctly, and solve the problem. This way, limitless people learn about the inventor, the product or brand, and the producing and distributing companies.

Then, in little or no time, the solution pervades and dominates the global marketplace for everyone who needs to acquire and use it. When this happens, you have demonstrated some of God's unique qualities.

Every globally used product and each universal brand attained its status via the same process. Somebody, a company, or a collaborating team meticulously executed research and invented the product. In direct or indirect collaboration with some multinational corporations, the invented solution is mass-produced and distributed worldwide.

People from everywhere ordered and used it repeatedly to solve the target problem and had the specific need met with ease every single time. Then, the inventor, producers, manufacturers and distributors became dominant in the global marketplace by domineeringly influencing people's thoughts and purchasing actions regarding the problem that the product solves.

When a product consistently meets and far exceeds the customers' expectations in excellently solving a persistent problem affecting the masses globally, it quickly dominates and sustains itself in the universal marketplace. Let your research discover, invent, or create such products. Then, in little or no time, the world knows and celebrates you as a prolific inventor.

When the world celebrates a person, his products and his brand, more customers get attracted to the products. This, in turn, increases the acquisition of the products. The sharp increment in a product's acquisition results in wider diffusion of the products and the brand. A worldwide success in an invention always leads to greater product success and the making of other products.

There is no better way for a person to demonstrate the unique characteristics of God in him than the creation of crucial values that effectively solve people's problems worldwide simultaneously who acquire and use the products. Invent irresistibly useful products that exhibit the extent of your all-powerfulness, all-knowingness, omnipresence, and creativeness.

BECOME A CLASSIC INVENTOR

Become a classic, paradigmatic and quintessential inventor. Be a perfect model of an inventor for people worldwide to emulate. Become a prolific inventor and the perfect model that the Invention Age needs.

You become a classic inventor when you fulfil your ultimate right, responsibility, obligation, and duty of making crucially useful inventions that are demanded or ordered from everywhere on Earth. By researching with every bit of your potential, resources and time, you make extremely useful inventions and create much value in the world. Then, you mass-produce, disseminate, and let the products diffuse and pervade the world. Consequently, you become a paradigmatic, quintessential, and perfect value-creator model for people to emulate.

As you keep creating value, you make it to the top one per cent of profound inventors. Then, you remain there as one of the greatest and most prolific inventors on the planet Earth.

As a result, people the world over, even those who began by antagonizing you by spreading deceitful and slanderous information about you and your products, will realize that they should rather emulate you and actually come to get mentored by you and learn from your exemplary products.

Yes, even the people who felt threatened by you and your products and responded first by throwing dust will come genuinely to receive mentorship from you. The people who, far from finding solutions to problems, chronically disparage and denigrate those who are investing everything in the pursuit of invention-making, will get changed and come to be mentored by you and learn from your quintessential products.

Be a maniac of invention-making. Make inventions and create great value until you get to and stabilize as part of the top one per cent of prolific inventors. Let the inventions you make positively affect humankind and propel countless people to accelerate faster to the maximum of their invention-making potential. Overwhelmingly, change people and the world for the best with your inventions.

Capture and sustain the world's attention with your inventions, discoveries, creations, and innovations. Get known everywhere by everyone with your products and brand. Become exceptionally popular with your inventions. Become the brightest light of the world, illuminating and dazzling people in the pursuit of invention-making.

Research, and research from time to time. Keep making inventions upon inventions. Continue to increase the inventions, discoveries, creations, and innovations you make even at the peak of economic depression. Even when the economy is nosediving to its lowest ebb, please continue to increase your research execution and the making of incredibly valuable inventions.

Never ever slow down, rest, or retreat for any reason, be it economic, political, social, health, technological, or scientific. No circumstance whatsoever should slow you down or prevent you from exerting extreme actions for the successful making of your target invention. Continue accelerating the exertion of unprecedented extreme actions for the exceptional making of more and many more inventions.

Take many more risks, the types that none else has taken or could take for making the target invention. Make a far more extreme investment of every resource you have in making the target invention, much more than other human beings have done or could do. Keep doing that committedly, dedicatedly, consistently, and persistently with passion until the actualization of each target invention.

Nothing, absolutely nothing, should retard the pace of your invention-making actions. Prove to the universe that you are unstoppably committed to the exertion of all the necessary extreme actions to make the target invention each time until an awesome actualization. Prove to the universe that you cannot retreat, slow down, or quit your pursuit of any target invention until it is wonderfully accomplished.

Generally, nature allows an invention to be made only when the researcher takes the type of exceptional risks that all others would not dare to take. It is when a person invests far more than what all other people could invest in the making of an invention that nature lets him actualize the invention. Please fiercely demonstrate your peak potential, capacity, energy, talent, and commitment to persist in the pursuit of an invention until it is excellently accomplished.

At each time, be completely committed to making an invention as it is your ultimate human right, utmost responsibility, greatest obligation, and absolute duty to invent and create value on Earth. Be readily prepared constantly for embarking on the making of exceptionally useful inventions. Only then can you successfully grab and work on the invention epiphany and opportunity that each persistent problem brings.

On completing each invention-making journey, let the product completely thrill the global audience, the customers. The product should capture and sustain people's attention worldwide by over-satisfying every user.

The over-satisfaction your product creates will completely capture the enthusiasm and commitment of the masses everywhere to acquire and use the product. The product of your research should so enthral the world and humankind like none else has ever done.

Each day, do something new for which you will be known worldwide forever. Take extreme actions daily that impetuously accelerate you closer to attaining the invention you are pursuing until it is excellently achieved.

Exert extreme actions daily that significantly move you forcefully and swiftly with impulsive vehemence and passion toward the exceptional actualization of your target invention. Keep doing that persistently until the stupendous materialization of the target invention.

CHAPTER 3

IMPACT THE WORLD WITH YOUR PRODUCTS

Positively impact the world with your products. Rise above everyone to be incredibly useful to all nations and people. The surest means by which an individual rises above all else and becomes most useful to people in all the nations on Earth is by making an invention of great value to humankind.
1. Invent products of indispensable worth to humankind.
2. Make your products universal.
3. Dominate the marketplace with your products.
4. Be an active participant in the global village.

INVENT PRODUCTS OF INDISPENSABLE WORTH TO HUMANKIND

Invent something, a product, solution, or service of indispensable worth and usefulness to humankind. It is only then that you will be of great value to people on all the continents of the planet.

Research and create globally indispensable value. Then, you become a greatly demanded brand and a global hero for inventing an essential product, service, or solution.

Regardless of how much money, acquired assets and possessions a person has, his ultimate satisfaction and fulfilment in life can, at best, be determined by how much he has positively impacted people worldwide with the value of the inventions he has made or funded others to accomplish. A person's life mission is never achieved until he makes or funds the making of useful inventions, creations, discoveries, or innovations for the best good of people everywhere on the planet.

A person's peak potential is only attained when he makes globally impactful inventions, discoveries, creations, or innovations of immense value to humankind. Create irresistible value with your inventions. Please, never leave your ultimate life mission unfulfilled. Do not leave your peak potential unattained. Research meticulously or fund others to make highly useful inventions to accomplish your life mission and attain your maximum potential.

The summit of true greatness in any field or profession is scaled only by the very rare individuals who research committedly, enthusiastically, courageously, and exceptionally accomplish their target inventions. It is by attaining the inventions a person's research targets that he can positively impact people and the world.

Please, be undaunted by pain or danger. Be a person of great courage and self-discipline to actualize the set invention goal in every research journey you embark upon.

Study this book and other similar ones by the current author to fully develop yourself for making inventions. Learn and grow yourself into a highly inventive authority with a thriving sense of self-worth and well-ingrained habits of goal setting and superb accomplishment of the goals – exceptional attainment of the target inventions.

Help people worldwide with your products to quickly reach the peak of their potential. For instance, the research and statistics books by Peter James Kpolovie alone have remarkably impacted people and the world. The books have exceptionally helped to bring invention making a foremost priority in business, industry, and lifestyle in the world.

His books have made people everywhere on the planet accept and treat the making of inventions as the ultimate human right, mission, duty, obligation and responsibility that should be individually or collaboratively fulfilled. The books have motivated people to execute research with greater rigour, triumphantly make crucial inventions, and create great value for humankind.

The books have become among the best, most celebrated, and most demanded books of that genre worldwide. The books made the making of inventions to become a passport to greater prosperity. The making of inventions has now moved to the front burner worldwide.

Invention-making has provided people with financial security and more value for prosperous family living. Almost every family now craves the making of inventions and exerts the needed action to invent something and create value in the world. It has become a thing of greatest pride for a family to create value, make inventions, and have a unique family product mass-produced and commercialized globally. Inventing a product and creating value has become a special feature that distinguishes a prosperous family from the others.

Suppose you acquire the books and make an invention. In that case, it implies that you are in the company of great kings and queens. You have joined profound and prolific inventors who improve people and advance the world. Invention-making has given uncommon prestige to each family that invented a thing.

As the books popularized the making of inventions, the world witnessed a viral boom in invention-making. Each person, family, group of people, company, multinational corporation, and state saw a greater need to make valuable inventions rather than merely acquire and use invented things. Factories and companies for mass manufacturing and distribution of invented products increased and became nearly ubiquitous.

Universities began awarding degrees based on the products the students invented instead of just theoretical academic programmes. Now, the academic programmes in some tertiary institutions of learning could only be said to have been completed by a student with evidence of invention made.

With the books, the number of inventions made in each state, nation, or continent spiralled up to the greatest heights annually over the years. The books brought more inventions to the world as many more individuals made inventions with their research.

Research came to be practically demonstrated universally as turning impossibilities into possibilities and unsolvable problems into solvable ones (Peter James Kpolovie, 2023). Each person embarking on research realized he was pursuing an unfailing quest for new possibilities. And exerted all the extreme actions to successfully arrive at amazing new possibilities. Countless people accelerated to the zenith of their potential by making inventions and creating great value with their research.

Individuals, families, collaborative teams, companies, and multinational corporations started accumulating and taking glory in their inventions rather than the hitherto accumulation of assets and possessions. They stopped investing in the accumulation of possessions not directly aimed at making inventions. With the books, people developed themselves fully for the making of inventions. They embarked on making inventions and accomplished the ultimate goal each time.

The books by Peter James Kpolovie have permanently made dominant marks on the minds of people everywhere on Earth that:

1. True prosperity, success, happiness and life-fulfillment are completely tied to the worth and volume of inventions they make.
2. The remarkably more that people want from life can only be attained with what they invent.
3. The greatest priceless aspect of life is the making of inventions.
4. Ultimately, a person can only attain peak prosperity, wealth and self-fulfilment when he makes valuable inventions, discoveries, and innovations.
5. Value creation and the making of inventions are the only things more to wealth, prosperity, and happiness than financial assets and acquired possessions.
6. Peak satisfaction lies in the inventions a person makes.
7. Invention-making is the utmost human right, mission, duty, obligation, and responsibility one must fulfil.
8. We live in the Invention Age, the era where what matters the most are the inventions we make and the values we create.
9. The future we crave can only be actualized with the inventions we make today.
10. Wealth creation and prosperity attainment depend on the worth and volume of value a person creates for humankind with his inventions.
11. Everyone who truly cares can research and invent by doing all the needful.

Consequently, individuals, corporations, states, and nations began intensely craving the making of inventions more than everything else. Individuals, families, corporations, states, and nations began investing in whatever it took to make inventions to meet the cravings. The results are visible with the inexhaustible inventions that pervade the world.

This is the Invention Age. Nothing can satisfy a person as much as the inventions he makes. As a result, embarking on and achieving the making of the target inventions has become the prime preoccupation of individuals, families, groups, multinational corporations, states, and nations. Influenced by Kpolovie's books, people have embarked on and made boundless inventions to attain peak satisfaction. The world is getting saturated with very highly useful inventions.

The books by Kpolovie make it abundantly clear that the Invention Age we live in demands everyone to create and disseminate great value. They emphasize that nothing makes life as precious and fulfilling as the inventions that a person makes or the values he creates. An individual's life will be fulfilling and precious to the magnitude of the value, worth, and number of his inventions, discoveries, creations, and innovations.

Like the prolific inventors on Earth, you can and should achieve everything you set your heart and mind on inventing. Just set that thing as the ultimate goal of your research. Execute the research committedly with everything in and about you.

Keep working on the research daily from strength to greater strength, nonstop, until it is excellently accomplished. Next, in collaboration with multinational corporations, mass-produce and disseminate the essential product or solution extensively to positively impact people everywhere daily.

MAKE YOUR PRODUCTS UNIVERSAL

Invent, mass-produce, and boundlessly disseminate each of your essentially valuable inventions or products in inexhaustible quantities until they pervade the universe. Let your inventions become present everywhere in the world simultaneously in inexhaustible quantities.

Having your invaluable products everywhere on the planet should turn your name into a most demanded brand across the globe. People from everywhere in the universe should know you and your products or inventions. They should acquire your inventions in multiple quantities per second, minute, hour, daily, weekly, monthly, and all through every year.

People should permanently think of you and desire to acquire your inventions or products ubiquitously. Let your inventions of extreme value make a permanent impact on the minds of people across the world. Your indispensable products, services, inventions, discoveries, ideas, and creations should be universally disseminated and demanded by people from all parts of the world.

Purchases made of and orders placed for your inventions should amass you all the wealth needed for making much more crucial inventions to help countless more people accelerate swiftly to their peak or maximum potential attainment. Invest the wealth your inventions amass in the improvement of living conditions for humankind. Touching and improving the lives of all humankind is the essence of making inventions and disseminating them globally.

Be everywhere at the same time with your inventions and positively impact the whole of humanity. Relentlessly research and make inventions that tremendously improve humankind to the extent that the world knows and celebrates you and your inventions. Let the world famously know who you are with your life-changing and world-improvement inventions.

DOMINATE THE MARKETPLACE WITH YOUR PRODUCTS

Penetrate and dominate the invention-making world with your extraordinarily useful solution, product, device, technology, idea, and service. Only then will the world know you and start paying dollars to benefit from the value you have created.

Break out of your obscurity by making an invention that the world greatly needs. Only a crucially useful invention you make for humankind will break you out of obscurity. When you break out of your obscurity, people from across the globe will look for you and pay the required amount of dollars to get your products, solutions, or services.

Mass-produce your product and disseminate it electronically across all continents on the planet. Let all those for whom the product was made be able to access and acquire it with payment of an easily affordable amount of dollars.

Take and rule over the global marketplace with your products. Become the expert and leader in the area of your invented product. Have and sustain dominion of the marketplace regarding your product. Take and rule over the space regarding your product.

Keep making more useful inventions, discoveries, and innovations repeatedly. And let it become obvious to humankind that your products have come to permanently dominate the marketplace. Let it be obvious to the world that you have developed to dominate as a prolific inventor.

Make everyone on Earth know you for and equate you with something, your product that people gladly pay to get. Let people world over equate you with the great value you have created on the planet. When the world knows you for something and equates you with the thing, an online search for that thing will automatically show or reveal your name every single time. Taking books alone, for instance, Peter James Kpolovie is known and celebrated for and equated with:

1. Research: Make impossibility possible.
2. Research, invent, and create wealth.
3. Excellent research methods.
4. IBM SPSS Statistics excellent guide.
5. Correlation, multiple regression and Three-way ANOVA.
6. Factor Analysis: Excellent guide with SPSS.
7. Multivariate Analysis of Variance: SPSS excellent guide.
8. Handbook of research for enhancing teacher education with advanced instructional technologies.
9. Statistical approaches in excellent research methods.
10. Educational management in developing economies.

Dominate and set the pace always in the business world with your products. Advertise and disseminate your products for them to dominate people's minds, thoughts and the marketplace. A great way for a person to advertise his invention is by making other related inventions. There is no better way for an individual or organization to advertise, disseminate and diffuse her product than to invent other related products.

With one product or invention, you cannot gain the full trust and confidence of people across the globe as much as when you make many inventions and products to better meet the needs of humankind. The greater and more the values you create, the more the world gets to know and trust you. The more the world knows and trusts you for the values you have created, the more dominant you, your products, and your brand become in the global marketplace.

Domination of the global marketplace is not for the majority who only depend on hope and prayers for their goal to be achieved. Far from hope and prayers, people's thoughts and the universal marketplace can only be dominated by making more valuable products or inventions to meet human needs.

Keep making more inventions, creating more values, and originating more products and solutions of greatest usefulness to humankind. Then, over time, you will dominate the global marketplace with your highly demanded valuable products.

BE AN ACTIVE PARTICIPANT IN THE GLOBAL VILLAGE

The world has since become a global village. Create indispensable value and become an active participant in the global village. It is only with your inventions that you can impact the world and benefit from the wealth therein.

Never ever rest on your laurels whenever you get an invention made. Keep exerting far more extraordinary actions and most successfully make more inventions until the exertion of extreme actions and making more inventions become your second nature.

The more inventions you make, the greater you invest in creating more amazing value. Every invention made naturally empowers, enables, and motivates the inventor to invest more in the triumphant making of far more inventions until he dominates the marketplace and the world with his inventions of immense worth.

Do not rest on your invention-making laurels. Keep investing in and making more extraordinarily useful inventions.

There is nothing good that cannot be invented. Every good thing imaginable can and should be invented. Be the person to get the invention made and made excellently.

When you invent a product, mass-produce it and take it to the international marketplace. Disseminate it effectively and efficiently until it dominates the world market.

Dominate the world market with your product. That is the purpose for which the Internet has made the entire world one

Global Village. Let people from all parts of each continent, across the globe, easily access your product and electronically place orders for it with a few buttons on the computer or handset.

Why has the world become a Global Village? The world became a global village chiefly to enable anyone who cares to create value, invent a product, discover a device, innovate something, and get it easily disseminated instantly worldwide.

Please do the following.
1. Invest extraordinarily in your research.
2. Create great value.
3. Invent crucially useful products.
4. Discover life-changing devices.
5. Make important innovations.
6. Mass-produce each of your products.
7. Take each of the products to the global marketplace.
8. Disseminate it everywhere on the planet.
9. Get paid worldwide for it electronically.
10. Positively impact every user and propel him to advance closer to his maximum potential.

Without creating value, disseminating it worldwide, and getting paid electronically for the value, a person has yet to impact the world. When a person creates value, mass-produces it, disseminates the value, and receives payment for it from across the globe, he has impacted the world.

It is with impacting the world that a person is, indeed, an active participant in the Global Village. Do everything necessary and become an active member and shareholder in the Global Village.

CHAPTER 4

BE RENOWNED FOR ONE THING – MAKING INVENTIONS

Be renowned for something in this world. The very best thing a person should be renowned for is making inventions. Be widely celebrated for making a crucial invention or creating a value necessary to humankind.

1. Be renowned for making inventions.
2. Concentrate on making inventions.
3. Set making an invention the ultimate research goal.
4. Motivation to action by ultimate research goal.

BE RENOWNED FOR MAKING INVENTIONS

For every research endeavour, set a compelling monumental life-changing goal. With consummate and outright concentration, focus on achieving the goal. Exert more than the mandatory effort to actualize the goal until it is fabulously accomplished. Then, become known across the globe for the indispensable value you have created.

Let your research concentrate on making an invention or discovery of tremendous value to humankind. Keep making inventions and discoveries with your research until you attain excellence and super-expertise in invention and discovery-making. Become a globally renowned inventor of indispensably valuable devices, things, solutions, services, ideas and products.

It is better incomparably to become an invention-making expert than in everything else. Nothing compares with concentrated research until a person becomes an excellent inventor.

Attaining excellence in invention-making is extraordinarily better than being good in other things put together. Singleheartedly focus and concentrate on researching and becoming a prolific inventor.

Strive consummately for excellence in invention-making rather than being good in everything else. Excellence in invention-making can attract every other good thing for the individual.

The same applies to companies, industries, multinational corporations, states or nations. This accounts for why every top-rated company or corporation globally is known for and thrives on the products, services, solutions and ideas it invented or discovered. Every globally renowned company is either commercially distributing its created products, or is rendering its unique services.

Readily give up everything else you love and concentrate on making inventions with your research. Focus completely on making inventions with your research until you become an expert inventor, a prolific inventor. Your inventions can attract every good thing to you.

There is nothing good that cannot be invented with enough rigorous research on it. Become the one who does the inventions that change human lives, improve society, and advance the world.

Research, invent and keep inventing different products to better people and society everywhere. Invent to make people become the very best they could be. Invent to perfect society and materialize the ideal world we crave.

BE RENOWNED FOR ONE THING – MAKING INVENTIONS

Nothing can impact and advance the world as much as the inventions we make with our research. Concentrate your life and everything you have on the making of invaluable inventions.

Do something of immense, indelible importance in your life. Do something that truly matters in satisfying the great needs of humankind and advancing the world.

Discover or invent an amazing product, an extremely valuable device with your research. A product that would be the greatest hit in the world when industrially mass-produced and commercialized.

Involve individual, corporate, and multinational investors in the mass production of the product. Engage a leading university, a booming advertising company and a most-viewed television station to design and float an international promotional campaign for the product.

Get the product sold to the public by the world's leading online stores. Next, the royalty accruing from the sales shall gain you more than the wealth and prosperity you need.

Change the world for the best with your inventions. Let nothing other than invention-making capture and sustain your attention.

Live for one thing. Be renowned and celebrated the world over for one thing. Invention-making is the only thing you should be famously known for.

Let your research results spark a scientific revolution. Your invention should spark a scientific revolution that subsequently involves hundreds, thousands and millions of scientists to fully understand all about the problem you investigated and the indelible solution you provided.

Be known all over as a foremost inventor. The attainment of that great feat demands every one of your thoughts, words and actions to be productively directed or targeted at making inventions.

CONCENTRATE ON MAKING INVENTIONS

Concentrate single-mindedly on invention-making with your research. Ensure that every one of your research journeys accomplishes the target invention or discovery.

Set an ultimate goal each time you want to embark on research. Concentrate everything about you on executing the research to achieve the set goal.

Complete concentration is the distinguishing characteristic, feature or trait of successful researchers, scientists or inventors. An invention can only be made with total concentration on the execution of the research for making that invention.

No one extraordinarily accomplishes research without total concentration of time, talent and treasure on achieving the research goal. Every exceptionally useful invention, creation, discovery or innovation that is the ultimate research target only gets attained when the investigator concentrates everything on achieving it.

A person, group of persons or corporate organization only make an invention or discovery of immense worth with total concentration on accomplishing the invention or discovery. With an unreserved concentration alone, every previously unsolvable problem in human history was solved.

Every hitherto impossibility in human existence was made possible due to one thing – passionate concentration, total concentration on turning possible the impossibility.

Please single-mindedly concentrate on making the invention or discovery your research aims to attain. Then, the invention, innovation, creation or discovery gets made, turning an impossibility into a possibility and an unsolvable problem into a solvable one.

Focused attention and total concentration on accomplishing the ultimate research goal in each research journey get the target invention marvelously achieved. Therefore, we should accord every research we embark upon the totality of our concentration until the attainment of the target invention, creation, discovery, or innovation.

For each research journey, clearly write out the ultimate research goal. Be convinced that the ultimate research goal is what can first be achieved by you and nobody else on earth. Next, make every needful sacrifice for, invest in, and exert all extraordinary actions on it until the superb actualization.

Keep the goal ever before you. Concentrate every effort on working for its achievement much more than any other human or group of persons has ever done or could ever do.

Let the ultimate research goal be compelling enough to enormously change people's lives, improve society and advance the world when accomplished. Concentrate and enthusiastically apply everything you have got to pursue the attainment of the goal.

Make all the investments in, sacrifices for, and take all the extreme actions that achieving the goal demands. Overcome all the challenges and hindrances separating you from actualizing the goal. Keep working on it with complete concentration without letting any distractions until the target invention or discovery is astoundingly made.

SET MAKING AN INVENTION THE ULTIMATE GOAL

Set the making of one major invention or discovery the ultimate goal each time you research. Focus dedicatedly on the awesome attainment of the ultimate goal. Do all the extraordinary work for its excellent completion. Plus, take every extreme action required to get it actualized amazingly. Intensify every effort and attention on attaining the goal until the invention gets made at all costs.

It is only by intensive working on attaining one research goal at a time that best achieves it. Set a clear invention goal you want to achieve with your research and do everything it takes to accomplish it outstandingly.

You can make an invention with a great goal that you focus everything in and about you on achieving with your research. Every inventor singleheartedly concentrated on making a specific invention before it was made.

Every invention gets done by having it as a goal and working concertedly on its actualization. Nothing can be invented by a person who is not dedicated to conducting research on making that invention.

Only a person, team, company or corporation committed to researching the making of a particular invention gets the invention made. Set out with your research for the making of an immensely useful invention. Single-mindedly work undistractedly on the research until the accomplishment of the invention.

Clearly set the making of an invention as your research goal. Remain focused with relentless nonstop action exertion on it. Dedicatedly follow it through to the excellent completion of the research. Only then that you achieve the invention.

Each research work is embarked upon for the attainment of a specific goal. There is no research execution without a specific goal. Execution of research must be aimed at making a particular impossibility possible or an unsolvable problem solvable by discovering, creating or inventing the solution.

There can only be research execution with an invention, innovation, creation, or discovery as the ultimate goal. There is nothing like a generalist research work conducted without any valuable invention, creation, discovery or innovation as the target goal. Similarly, no single research work can solve all the problems of humankind.

Every research must have an important goal, a target invention, discovery, creation, or innovation of monumental worth.

There cannot be a generalist research work. A scientific quest that is aimed at nothing in particular accomplishes nothing. No one can hit an invention or discovery of value with a research endeavour not aimed at any specific product, solution or service.

Aim at nothing with your research, and you will surely:
- Invent nothing
- Create nothing
- Discover nothing
- Innovate nothing
- Accomplish nothing.

What a person aims at and does all the needful for its attainment is what he accomplishes. The person who aims at not accomplishing anything does achieves nothing with his research journey. He ends up nowhere and makes no invention.

Research can only be said to have been done when it accomplishes a goal. Without accomplishing a goal, research has yet to be done. Nothing can be empirically achieved with research when it does not aim to make an invention, innovation, or discovery of something of value.

Aptly, research must be executed to make possible an impossibility, solve an unsolvable problem, or bring into existence a nonexistent thing (something that has never existed before). Without making an invention, discovery, creation or innovation, research has not been done.

Every research must have a clear target. The research will only be said to have been done when the target was hit. The research is accomplished on attainment of the target creation, invention, innovation or discovery.

Never dilute or shift your attention to things other than the actualization of the research goal. Never hop from one research goal pursuit to another without first achieving it. The research goal is never attained by hopping from one target to another inconclusively. Dedicatedly pursue and accomplish one research goal at a time.

Let me reiterate the crucial points made. When a person aims at achieving nothing with his research, he sure ends up there. He invents nothing, creates nothing, discovers nothing, innovates nothing, and adds no value to society and the world.

The majority of the millions of people annually who claim to have attempted research without achieving any useful invention, creation, innovation, or discovery did so because they had yet to have any specific goal which they conclusively pursued. In research, without a useful goal, nothing worthwhile can be achieved.

When an individual aims at an insignificant, small, or valueless thing with his research, he puts in little effort and sacrifices little or nothing to execute the study. Consequently, he achieves nothing, or at best; he only attains something of very little or no value to people and society. He attains something petit, minor, irrelevant and almost useless.

Conversely, when with big aspiration and outright determination, a person sets out for a great and highly demanding goal with his research, the odds are that he exerts the requisite effort and invests all the necessary time, talent and resources to attain the goal. Consequently, he accomplishes a worthwhile invention, discovery, creation or innovation.

Having a specific incredibly great goal for a research work increases the investigator's craving, drive, and actions for its actualization. The ultimate research goal motivates the investigator to invest everything and sacrifice all his talent, treasure and time in pursuing the materialization of the goal until a wonderstruck actualization.

Committedly pursuing a great research goal culminates in making a major invention or discovery of tremendous value to people, society and the world.

When every effort is exerted to achieve a big and extremely useful research goal, it gets attained. Then, the invention or discovery improves people, perfects society, and advances the world by a giant stride.

Therefore, set out committedly with your research to accomplish a big, great and extremely useful invention, creation, discovery, or innovation. Constantly remain focused with every needful action for accomplishment until it is actualized.

Be completely dominated by the compelling thoughts and actions for making the extremely valuable discovery or invention the research aims at. Work dedicatedly nonstop on it until a wonderstruck attainment of the goal. Persistently work relentlessly on materializing the ultimate research goal until a thunderstruck accomplishment.

This is the only way to make life-changing and world-improvement inventions and discoveries. Aim big, exert every needful effort, take much more than necessary action, and sacrifice the required time, talent and treasure to achieve the ultimate research goal.

MOTIVATION TO ACTION BY ULTIMATE RESEARCH GOAL

Let the overall goal of your research, making an extremely useful invention, motivate you to take all the actions for its accomplishment. Be consistently motivated toward the excellent accomplishment of the invention until it is awesomely done. The ultimate research goal should consistently motivate and fuel your acceleration of greater and much greater effort until the dazzling actualization of the target invention, creation, discovery or innovation.

Work harder and smarter than anybody has ever done to make your target invention. Prioritize ever-increasing effort exertion in the excellent completion of the invention-making journey until the exceptional attainment of the target invention.

Remain resolutely enthralled in the exceptional attainment of your ultimate research goal. Always have a crystal visualization of the rewards that making the invention will bring, even when they are long away and when there are no guarantees of being rewarded.

Have and demonstrate an ultimate goal-oriented mindset. Translate your intention, ambition, potential and best talent into action. Only enthusiastic, ever-increasing actions will make your good fortune and fabulous invention dream a reality.

Take all the actions for the actualization of your ultimate research goal. Your consistently accelerated actions alone will materialize your intended inventions and discoveries. Exert the necessary action to achieve the exceptional research goal in every invention-making journey you embark upon.

It takes months and years of intense practice, hours a day, every day of the week, to learn and master the skills of research execution for unfailing invention-making. As a scientist, a person may have to do hundreds of experiments to get the one invention or result that makes an impossibility possible and turns an unsolvable problem solvable for the best good of all.

Dedicate an untold number of hours to committed tedious, repetitive practice that successful research execution and invention-making demand. Keep practising and improving your research execution skills to ensure exquisite attainment of your invention-making goal in each research work you embark upon.

CHAPTER 5

BE PHENOMENALLY GENEROUS TO INVENT

Be incredibly generous to invent. It is with incredible generosity that inventions are made. To invent, a person must be phenomenally generous in making all the necessary sacrifices for and every investment in the invention-making journey that the value creation demands. With phenomenal generosity alone, a person, collaborating team, company, multinational corporation, or nation invest excessively more than all necessary to make an invention of exceptional value for humankind.

1. Be incredibly generous to invent.
2. Generous use of talent, treasure and time.
3. Largesse over parsimony.

BE INCREDIBLY GENEROUS TO INVENT

With incredible generosity alone inventions are made. To make an invention, a person must be incredibly and limitlessly generous. Only phenomenally generous persons do herculean things to make inventions that best improve society, better people and advance the world.

Every invention made has an expansive effect on the inventor and the public. Each invention magnifies the inventor and expands his contributions to the world's advancement.

Each research work executed and every invention or discovery a person makes, build up his experience and stamina that better prepare him for accomplishing greater and more breathtaking inventions for swifter advancement of the world. Every successful invention making is a veritable aircraft for realizing a more expansive goal of rendering positive service in the world's advancement.

Nature offers an infinite abundance of resources in this world to enrich those who have cravings and the willpower to make inventions. The resources enable them to pursue their dreams, aspirations, visions, goals, and cravings to marvellously triumphant ends. Whatever invention you exceptionally crave and set out with a complete commitment to make with your research shall get done once you do all the needful.

Though resources and opportunities for invention-making are limitless, a person must work fast on every invention-making journey, get it made and protect it with a patent or, at least, a copyright before any other person does it. When you get the invention made, it helps and facilitates others to pursue the completion of their invention-making journeys and the actualization of their peak potential.

Each successful invention creates massive opportunities and encourages, spurs and propels countless people worldwide to accelerate swifter toward their apex potential. Each excellent invention accomplished is like a multinational corporation that employs hundreds or thousands of people to work to attain their individual and collective goals.

Most importantly, each invention-making goal materialized also attracts like-minded people, investors, and companies with similar and related interests or goals to the inventor. Such attractiveness enables the inventor to easily actualize his sense of expansiveness and willingness to positively impact and affect nations across all continents.

This accounts for why making an invention demands the investigator or researcher to be phenomenally generous. It may take exertion of untold hard work, passion, enthusiasm, persistence, and the investment of everything he has into the invention-making journey before a universally valuable product could be invented. Once made, he gains back everything in a million folds and dominates the world.

Only with phenomenal generosity can a person make all the investments and sacrifices that the making of a globally essential invention demands. Be overwhelmingly generous to invent. Then, the invention shall attract like-minds, investors, companies and multinational corporations with related interests to you and simultaneously actualize your expansiveness in positively affecting the approximately nine billion people in the world by 2034.

The investors your invention attracts shall enable you to most successfully embark on and accomplish much more valuable and demanding inventions. This is how the making of one invention often leads to the making of much more valuable inventions.

Please, start and get one invention of great value made. Next, you will experience how that invention galvanizes you into making many more essential inventions.

GENEROUS USE OF TALENT, TREASURE AND TIME

Use your talent, resources, and time generously to make inventions, discoveries, and creations. A person loses what he does not use. Whatever you fail to use, you lose.

Countless people have lived and departed from this world without making any invention, creation, discovery, or innovation. They never invested their talent, time and treasure in value creation while on Earth. So, they lost them forever.

Invest what you have and are generously in making crucially useful inventions and creating exceptional value for humankind while on Earth. The inventions a person makes are the greatest gifts he could present to humankind for the continuous advancement of the world.

The development of humankind and the world's advancement shall be stagnated, stalled, and retarded if we do not generously invest our talent, time, and treasure in making useful inventions. Hoarded wealth, resources, time, talent, energy, and effort cut off the dynamic flow of life, improvement, progression, and advancement of humanity and the world.

Refrain from hoarding your energy, effort, resources, time, talent and treasure from investing heavily in successful value creation. Invest them liberally in making inventions. The progress, development, improvement, and advancement of people, society, and the world depend solely on the value we create and the inventions we make with our research.

Our attainment of prosperity and self-fulfilment depends upon the inventions, discoveries, creations, and innovations we make with our research. Please, wholeheartedly invest all you are and have in making inventions of great worth. It is the best way a person can attract, amass, and sustain wealth in abundance.

Only when you make inventions that you have not totally lost or waisted your talent, special gifts, time, resources, energy, and treasure while on Earth. Invest in and achieve new possibilities and novel solutions for permanently solving persistent human problems. Inventions are the very best gifts a person can and should contribute to the sustenance, spread, extension, and preservation of God's creative work.

The invention you give to the world multiplies and yields you uncommon prosperity in return. Create value for the world and, in return, enjoy the wealth, prosperity, self-actualization and self-fulfilment it brings.

Everyone who wants to get the best in life creates and adds value to the world. Only the value a person gives the world that attracts him the best of everything the world offers.

Create great value with your research. In return, you get the best from nature and the universe. The world's wealth only flows to and remains with the individuals who have created astronomical value.

Perpetuate your existence on Earth with the permanent inventions you make and the indelibly monumental value you create. Add mammoth value to human lives, society and the world via the research you execute.

The eternal value you create for the world with your inventions is the best way to perpetuate your existence. With the indelible value you create, the world endlessly remembers, honours and celebrates you. You become an all-time hero on the planet with the value you create.

The value and inventions you give to the world maximally increase your sense of prosperity, wealth, life mission accomplishment and self-fulfilment. Never fail to add perpetual value and essential inventions to the world.

Make inventions that help others become more capable, independent, and better accelerate swiftly to attaining their maximum potential. Let your inventions move people worldwide to the actualization of their zenith potential.

With your research, make crucial inventions and create amazing value that positively affects people everywhere on Earth. The more people who strive desperately to attain the peak of their potential as influenced by your products, services, ideas, and solutions, the better the world becomes, and the more prosperity you experience.

LARGESSE OVER PARSIMONY

Largesse is the generosity with which a person invests his talent, time and treasure in research and mass-produces the invention made for best dissemination to the global audience. It is with largesse a person passionately invests in executing research for the creation, invention or discovery of exceedingly valuable solutions, products, services, ideas, technologies, and knowledge for the best advancement of humankind and the world. Largesse leads to creations and inventions.

Largesse is the opposite of parsimony. Parsimony never leads to the making of exceptionally useful inventions.

Parsimonious individuals are principally motivated to look for and do only the easiest, cheapest, simplest, least challenging, closest, and riskless things. Parsimonious people are too selfish, greedy, miserly, and niggardly to invest energy, time, resources, or talent in seeking functional, novel, new and uncommon solutions to persistent pressing problems.

Persons with parsimony typically value what currently exists and what they have over researching to make inventions of greater value. Parsimonious individuals prefer conserving what they have instead of investing in research execution for successful invention, discovery, innovation, or creation of things of extreme value to all or the majority.

Parsimonious people are never eager to invest in making inventions that will alter, change or improve the status quo. They never research to invent what will alter the existing convention or status quo.

People predominantly characterized with parsimony hold fast to what they already know and keep doing exactly what they do. They intentionally block their brain or memory from restructuring and learning new things. They are resistant to any change that will greatly advance human development.

Parsimonious persons are characteristically motivated by convention to be lazy and to do only what is convenient. They are never willing to engage in carrying out research chiefly because the journey of making inventions and discoveries is hard, very demanding and highly inconvenient.

Parsimonious people are niggardly and typically gravitate toward not executing research. They refuse to research because not researching is an option that requires the least amount of effort or work. They are more pleased with living with persistent problems instead of investing in research to create, invent, or discover lasting solutions to the problems.

The people habitually dominated by parsimony are, by convention, motivated to do what is easy. Doing only what is easy is the option the masses take, and it does not culminate in invention-making.

It is not the masses that make inventions. The masses do not create or discover things of greatest value to all. The predominantly parsimonious masses do not engage in making innovations of immense value for the advancement of the world.

The parsimonious people are only concerned with consumption or using up the creations and inventions of other people. Please research. Do not be part of the masses, the parsimonious people.

It is much easier to merely use up invented things than to research and make more valuable inventions. Parsimonious people constitute the masses that prefer to merely use things invented by others instead of inventing much better things for humankind.

A parsimonious person would prefer being lazy and doing only what is convenient. Laziness and exertion of convenient actions are things that everybody can do. Doing whatever everybody does or can do never results in the making of inventions.

A person who does not use his brain to think of creatively

solving unsolvable problems, never gets anything invented that makes possible an impossibility. Someone who does not do the hard things others refuse to do never invents or discovers an uncommon solution to a persistent problem.

The making of inventions, discoveries and innovations of immense value is only achieved by largesse people. Only people who think and do things extraordinarily for the best good of all or the majority are those who create, invent, discover, or innovate with their research.

Creation, invention, discovery, or innovation of things of great value demands a committed investment of time, talent and treasure in doing the hard things that all others never attempt to do. Only the largesse people rigorously research and actualize their target inventions.

On the contrary, parsimonious people conserve their energy, time and resources. They never use their brain generously to find solutions to unsolvable problems.

Consequently, the parsimonious people merely live in and depart from the world without making any indelible invention, creation, discovery or innovation. People who do only the easiest, cheapest, and most effortless things that everybody else does or can do never make any inventions.

Please, do not become like the parsimonious people who only use the products of others' research without embarking on research to make much more valuable inventions. Embark on research and make inventions and discoveries that add much greater value to humanity and the world.

Routinely gravitating toward the option requiring the least effort is equivalent to following the crowd. It does not lead to the making of any worthwhile invention. Doing only the simplest things, like everyone else, does not result in any remarkable accomplishment.

Research works executed successfully for making essential inventions never followed the simplest course of action. Kindly get actively and dedicatedly engaged in executing the research that all others will not even dare or attempt to do. Research and create, invent, or discover solutions to unsolvable problems of humankind.

Researching and making valuable inventions are only done by largesse people. Largesse people willingly invest their unreserved effort, resources, time and talents passionately and selflessly for the best good of all. Generous people who develop and invest their brains, exert their energy, and hold back nothing in researching are the only people who make inventions. They are the people who progress society and advance the world.

With largesse, a philanthropic and liberality of heart, let your research do the hard things now that pay off in the long run for the best good of all. With your research, do only the very hard things that others cannot do and will not dare to do. The hard way is the only way that leads to making inventions excellently.

Never do the easy things everybody else is doing or can do with little effort. Please, make the greatest investments in, and sacrifices for, the invention of things of indelible worth for the best good of humankind and the world's advancement.

Easy things that everybody can do, such as devoting many hours to social media and television watching, scrolling on the phone, and email checking, never leads to the making of inventions, discoveries and innovations of great value. Doing what everyone is doing or can do hinders innovation and invention-making.

Following the masses never changes the status quo for the best. Refusal to do research that stands out never produces anything of tremendous worth for the advancement of the world.

Therefore, to uniquely improve people's lives, remarkably progress society and advance the world, a person should rigorously embark on research, passionately work on it committedly and achieve the essential invention the research targets. Alternatively, a person should sponsor others to execute such research.

Making extremely useful inventions, creations, discoveries and innovations with one's research works until the person becomes a prolific inventor is the hard thing that only very few people who are characteristically dominated with generosity, largesse and unselfishness do. Please be extravagant with researching, inventing, making impossibilities possible, and turning unsolvable problems into solvable ones for the best good of humankind.

CHAPTER 6

STRIVE FOR AND ATTAIN ABUNDANCE BY INVENTING

Strive for and attain abundance with your research. A person or an organization attains abundance when her research invents products that dominate the global marketplace in the chosen field.

1. Invent and attain abundance.
2. Achievement of true greatness by inventing.
3. Attain success with making inventions.

INVENT AND ATTAIN ABUNDANCE

Research and create one great value, another product of immense worth, and another invention of extreme usefulness to humankind. When you keep making such inventions, the products shall very soon dominate the global marketplace in your chosen profession. Then, you attain abundance.

A person attains abundance to the point that whenever people across the globe think of the profession, they automatically think of the person's products and the indispensable value he has added to the world. A person's inventions dominate the global marketplace when his profession cannot be thought of without automatically thinking of the unique, indelible products he has invented, created, or discovered.

When throughout the world, a field of human endeavour can much better be practised with the application of the breakthroughs a person has made, then he has attained abundance. That is, abundance is reached when a person's products, created values, and the scientific cum technological breakthroughs he has made dominate a profession.

A person attains abundance when his unique contributions to his field of professional praxis become indispensable for the sustenance and advancement of the profession. Kindly research and create value that dominates the marketplace and magnetizes an abundance of wealth to you.

Create value upon values until you become part of the 20 per cent of inventors whose inventions and values created are worth more than 80 per cent of all inventions and values created by all others on the planet. Yes. You can get there by consistently making one invention of great value after another until your inventions dominate the thoughts of humankind. Then, you achieve super-abundance as one of the 20 per cent of inventors who have invented more than what 80 per cent of all other inventors on Earth have collectively done.

To achieve dominance of the world marketplace and magnetize super-abundant wealth with your inventions and created values, you must practice and over-master the skills and characteristics for invention-making as emphasized in this book, **Research, Invent, and Create Wealth,** and in two earlier books, **Research: Make Impossibility Possible**, and **Excellent Research Methods**, by the current author (Peter James Kpolovie, 2023; 2016). The others of his books are also necessary resource materials.

Attainment of dominance in the global marketplace demands breaking out of the average class, normal class, middle class, or status quo class in your research execution. You have to execute research extraordinarily well and fantastically actualize the ultimate goal in every one of your invention-making journeys.

In each research journey, you must set an exceptionally fine goal and exert extreme actions with relentless enthusiasm until a thunderstruck actualization. You must invest exceedingly more than the normal or average creativity, talent, effort, energy, time, and treasure in each invention-making journey.

Each time you embark on an invention-making journey, you have to invest everything in and about yourself in the unfailing pursuit of the ultimate research goal, exceedingly more than what 80 per cent of other humans would and could do to attain the target invention.

The people who make average or normal investments of their time, talent, potential, and treasure in the execution of research have yet to get it done. They do things the way almost everyone else does. A person only makes the invention when he thinks and exerts actions differently from all others during the research journey.

Embarking on research like everyone else does; never takes someone higher than the middle or average class level who never accomplishes any invention. The people who put in only the normal or average sacrifice, effort, creativity, energy, talent, time, and resources in the invention-making journey never invent anything because they do not think and act exceptionally as demanded by the making of inventions.

Break out of what I call the "status quo mentality" that does not require you to exert extreme action for invention-making and value creation. The "status quo mentality" is satisfied with things as they are and will not compel enormous investment in the pursuit of invention, discovery, creation, or innovation that is currently unthinkable, unsolvable and impossible.

Unfortunately, over 99 per cent of all those who claim to embark on research only work within the "status quo mentality" and end up inventing nothing, discovering nothing, creating nothing, and making no life-changing innovation. Those with the "status quo mentality" do not get research done. The people with the "status quo mentality" never complete research. They do not invent anything. Never operate with the "status quo mentality". Desperately crave, strive for, and achieve abundance with your invention-making journeys.

ACHIEVEMENT OF TRUE GREATNESS BY INVENTING

True greatness is best achieved with the making of inventions. Making an invention, discovery, innovation, or creation of tremendous value to humankind gives a person the feeling of peak success attainment. Accomplishing different things can make an individual experience the feeling of being great and successful. Attainment of different goals could depict greatness and success to different people at varying life stages, situations, conditions, periods, and events.

The things that a person accomplishes and experiences real greatness and success will not be considered successful by several other people. Just as great goals people can pursue are countless, so are the things somebody can accomplish and experience real greatness and success.

But there is only one thing: whoever accomplishes it attains peak or maximum success and greatness from all perspectives. Every person who achieves only one thing attains maximum greatness and success, irrespective of his goals, situations, conditions, or stages in life.

That singular thing that could give everyone maximum success when he accomplishes it is making an invention, creation, or discovery of great value to humankind.

Every person who makes an invention, discovery, creation, or innovation of tremendous usefulness to humankind will experience real peak success and greatness. Everyone who makes an invention of superb worth will, by it, attain peak success and greatness.

Every person who creates, discovers, innovates, or invents a thing of immense usefulness to humankind personally celebrates or experiences success and humanity celebrates him as a successful person. When a person makes an invention, creation, discovery, or innovation that advances the course of humanity, he leaves an enduring legacy on the planet and is timelessly remembered as such.

Who will not experience peak success if he makes an indelibly useful invention, discovery, creation, or innovation? No one. Every person who creates, discovers, or invents a thing of essential value achieves great success.

Therefore, since we value attaining real peak success and greatness more than everything else, we should dedicatedly seek the making of inventions far and above everything else. We should wholeheartedly invest everything else in the pursuit of invention-making until astounding completion. Yes, let us invest the best of our time, resources and talent in making useful creations, discoveries, innovations, or inventions until outstanding accomplishment.

Human life on Earth is a journey aimed ultimately at making indelibly useful inventions. The only people who accomplish their life's mission of creating something of tremendous value are those who always pursue the fulfilment of that goal with the mindset of winning it at all costs. Only the people who, at every moment, demonstrate with their thoughts and actions that they are on this planet to make important inventions, regardless of whatever it takes, get the inventions made.

Please, practically demonstrate the mindset that 'I am on earth to win the making of crucially useful inventions.' The making of each invention results from relentless, appropriate, extreme actions taken over time to actualize a truly unique vision or goal that advances the world. An invention can only be made with ceaseless extreme exertion of the right actions until the awesome actualization of the target invention. Without relentless exertion of all the necessary actions, no matter what, a person does not invent anything.

ATTAIN SUCCESS WITH MAKING INVENTIONS

Attain success, real success, with the making of inventions. It is only by making at least one invention that a person achieves genuine success in life. The attainment of success by an individual, a team, an organization, or a nation depends upon the volume and worth of inventions she makes.

Success via invention-making is everything. Nothing gives a person practical experience of success as the making of an invention of great worth. The more inventions a person, organization, state, or nation makes, the more successful the individual or establishment becomes in advancing the world.

Success in invention-making is absolutely crucial in the life of a person, group, corporation, society, state or nation, and the world, regardless of the goals pursued. Success in making each invention improves everything across individuals, society, and the globe.

The completion of an invention-making journey optimizes a person's sense of self-fulfilment. When somebody makes an invention, he feels the peak sense of having made a significant contribution to the advancement of people, society and the world. The invention (the product invented) gives the inventor the highest sense of success, confidence, security, and wealth.

Nothing benefits humankind as the making of an invention,

discovery, creation, or innovation. Making an invention benefits everyone and every sector of the economy. When an idea capable of improving people, bettering society, enhancing everything, and advancing the world is not invented or created, nobody and nothing benefits from it.

An unachieved invention goal never benefits any person. It never improves a thing. And it does not progress the world any bit. It never improves the world. The unactualized invention-making dream never advances the world.

Only the inventions made contribute to people's betterment, improvement of society, and the world's advancement. An unwritten and unpublished book cannot be of use to people. An invention epiphany that is not researched and materialized cannot be useful to people or the world.

Individuals, groups, society, and the world only improve or advance with inventions made and books written and published. Only what is done adds to people's success. What is done alone that counts in the betterment of the world.

The invention-making vision not acted upon and materialized contributes nothing to people's betterment, society's and the world's advancement. An intended goal of creating value or making an invention only improves people, enhances society and advances the world when it is pursued to conclusive actualization. Please, exert all the extreme actions to actualize your target invention and achieve peak success.

The survival and advancement of people, society and the world depend on the worth and volume of inventions made. Individuals, groups, organizations, states, and nations must actively work and achieve their invention goals to remain in existence and advance. Without making inventions, individuals, society, corporations, states, nations, and the world will stagnate, retrograde, and cease to exist gradually until it eventually comes to nothingness. What sustains us and the world are the inventions we make.

Every thriving company, industry, multinational corporation, or nation must continue to invent, create, discover, or innovate products, commercially mass produce, and distribute the products to satisfy the needs of the clients, customers, employees, investors, and other stakeholders. With such products and services, society and the world progress and advance. The betterment of the future depends mainly on the products and services continuously invented and disseminated commercially.

The world's sustainability in an ever-improving form is a function of important or vital invention-making goals attained. Continuous sustainability and advancement of human lives, society and the world depend upon the values we create, invent, or discover with our research today.

It is our individual and collective responsibility not only to sustain, but to swiftly advance the world by making vital inventions, discoveries, creations, and innovations. These, we can do only through our research.

Our greatest duty, responsibility and obligation is to make essential inventions, discoveries, and innovations for the sustainability and advancement of society and the world. Making highly useful inventions is an ethical duty each of us has to accomplish at present for a far better future.

Making crucial inventions with our research is the surest means for attaining real success, greatness, glory, and wealth. We have got to make all the sacrifices for, and investments in the making of inventions to best meet the great needs of humankind. To attain a life of abundance, a person must invent and create great value in this world. Please do everything it takes to make indispensable inventions for the astronomical improvement of humanity and the advancement of the world.

CHAPTER 7

UTMOST HUMAN RIGHT IS THE MAKING OF INVENTIONS

The making of inventions is the ultimate human right. We should individually embark on the making of an invention as an absolutely necessary human right, mission, obligation, duty, and responsibility that must be accomplished.
1. Making inventions is my ultimate human right.
2. Infinity of things to invent.

MAKING INVENTIONS IS MY ULTIMATE HUMAN RIGHT

Embarking on an invention-making journey nonstop to fruition should not be taken as a mere choice. Rather, it should be treated as a must.

We should each embark on making an invention as an ethical and honourable human right, duty, responsibility, and obligation of topmost priority until the exceptional achievement of the target invention. It is an absolute must for a person to create and add an unforgettable value to the world with his research.

The loftiest and utmost fundamental human right is the making of inventions. We should invest 50 per cent more than all it takes to make an invention individually. Making inventions is our top lofty fundamental human right. A person should embark on making an invention as it is the pursuit of his most exalted fundamental human right.

It is the peak of our individual human right to outstandingly make an invention of tremendous value. Let us each pursue the making of crucially important inventions as his ultimate human right, duty, responsibility, and obligation. It is our most essential right to individually have an invention of great value made.

Meticulously research from time to time and make extremely useful inventions until making inventions becomes your dominant habit. Habituate the making of inventions. Repeatedly complete invention-making journeys until it becomes your top habit. Completely commit yourself to making inventions via successful research execution from time to time.

With gusto, commit yourself to completing each of your invention-making journeys exceptionally well. Embark on and complete the making of inventions, one after another.

The making of inventions is our individual and collective most fundamental right, ethical and moral duty, responsibility, and obligation to humankind. The sustainability of the universe and the advancement of the world depend absolutely upon the magnitude to which we fulfil our most fundamental human right, ethical and moral duty, obligation, and responsibility of making inventions of great value.

It is our topmost fundamental, moral and ethical right, duty, obligation, and responsibility to take much more than the necessary extreme actions to reach our maximum potential by individually and collaboratively making essential inventions. Making inventions is our supreme fundamental human right, responsibility, obligation, and duty. It is our utmost legal and ethical human right, mission, duty, obligation, and responsibility to make inventions, discoveries, and innovations that satisfy human needs and engender a much more blissful future.

The future we crave most is in our own very hands to create by making all the requisite inventions it demands.

We must individually, and sometimes collaboratively, live up to our utmost human responsibility, obligation, duty, mission, and right to make valuable inventions.

Our **utmost human right** which is **the making of inventions**, is something so marvellously excellent that before the publication of this book, the socialization process in the Constitution of the United States of America, and those of other nations, never included or emphasized. Henceforth, **the creation of value** as our **ultimate human right** must be prioritized and accentuated in the socialization process until every individual learns it, realizes it, craves it, and does everything necessary for its actualization.

Our utmost human right is to make useful inventions and create value on Earth. The greatest and most important responsibility, obligation and duty we individually have to fulfil is the making of crucially useful invention. The most fundamental human right, greatest obligation, responsibility and duty we individually owe humankind is the making of useful inventions and creation of great value for improving lives and advancing the world.

To make the invention, we must individually exert extensive action each day as though it is the only day we have to complete the target invention. Extremely invest everything daily in the fantastic completion of the invention with which the world will indelibly know and idolize you. The magnitude of value a person creates is proportional to the extent to which the world knows and idolizes him.

The most pernicious evil or damage a person can do to himself, his society, state, nation, humankind, and the world is discarding his utmost human right to make inventions and create value. Never fail in your life's ultimate right, mission, duty, obligation and responsibility of creating great value and making crucial inventions.

The person who makes no invention and creates no value is a regret and a practical curse to humanity. A man who contributes no value to the sustenance and advancement of the world is an irreversible curse to humankind because he refuses to live up to his utmost human right, responsibility, obligation, and duty.

Please never be a curse to humanity. It is only when you invent something useful and create value that you cease from being a curse to humankind.

On average, the established tradition does not expect or demand a person to make a crucial invention. That is why in our upbringing before the publication of this book, value creation and the making of essential inventions was not emphasized as our ultimate human right, responsibility, obligation, and duty.

Fortunately, as you read this, you have realized that creating great value and making crucial inventions is an unconditional right, responsibility, obligation, and duty you have to fulfil. Please, live up to it. It is the very best way to distinguish yourself from the rest of humankind and other creatures.

Living an average life never allows for the making of an indelible invention. You and I are, therefore, charged to break out of the hypnotic state of mediocrity that an average life compels people. We have got to live an extraordinary life. Living an extraordinary life enables us to invent and create great value for the sustenance and advancement of the world.

We must individually and collaboratively make a constant, relentless effort by exerting colossal action to make a world-changing invention. Make all the investments in and take all the committed monumental actions that the timely completion of an essentially valuable invention for the improvement of people, betterment of society, and advancement of the world demands.

It is our absolute individual right, responsibility, obligation and duty to make an invention and add great value to the world.

Nobody else will make your sublime dream of creating great value or making a crucial invention come true. The onus of making a highly valuable invention and creating compelling value rests solely on each of us.

To stand out from the crowd, we must individually or collaboratively invent, create, discover, or innovate a product of great value, mass-produce it, and disseminate it for people worldwide to access, acquire and use. We have got to individually invest all that it takes, no matter what, to make an invention of tremendous value to humankind.

From this moment, make the great commitment to fulfilling your onerous right, mission, responsibility, obligation and duty of making an invention of immense worth by massively investing whatever it takes daily until the invention is excellently made. It is the best way to improve lives and society and advance the world.

Each of us has incredibly greater creativity, talent, energy, and potential for making inventions than what we are using. Only an extremely high level of investment in everything about us for the making of inventions can get a product of great importance made.

We individually have abundant creativity, energy, gifts, talent, or potential for the creation of incredibly valuable products, solutions, services, inventions, discoveries, or innovations that are yet untapped. Maximize your potential for making discoveries, inventions, or creations to radically accelerate the world's advancement.

When an individual maximally utilizes his invention-making potential, he can invent anything of great worth that he sets his mind on. Fully exercise your potential and extremely invest it in making inventions that best meet human needs than all that already exist. Take total responsibility and live up to your fundamental human right, duty and obligation of making inventions and creating values for the betterment of humankind and the world's advancement.

A person only makes an invention of immense worth to humanity by investing substantial talent, resources, treasure, time, and action. Very kindly make the extreme investment of everything about you in each of your invention-making journeys. Then, the target product of tremendous value shall excellently be attained every time you embark on an invention-making journey.

There is no option other than the extreme investment of a person's talent, resources, creativity, energy, action, and time to make astonishingly valuable inventions. To invent, create, innovate, or discover, make an extremely massive investment of everything you are and have in executing the research upon which making the invention depends.

Exceptionally enormous investment of a person's talent is necessary for successful invention-making. Exertion of extreme action is mandatory to attain a life-changing and world-improvement invention. Exceptionally huge sacrifice of one's comfort is indispensable for creating great value.

Never has a person invented, and never will someone, invent something of immense value without first exceptionally investing his potential enormously in the project. Please, committedly make such extreme investment of your time, talent, and resources or treasure in your invention-making journey. Then, the target invention shall unfailingly be achieved.

The more extraordinary action you exert toward attaining your invention goal, the better your chances of achieving the breakthrough. The more extreme your investment in pursuing your target invention, the greater the odds of fantastically achieving the target.

The more exceptionally monumental the personal comfort you sacrifice for making an amazingly valuable invention for humankind, the finer your chances are of excellently actualizing the target invention. The more you master research execution skills, the closer your target invention gets to you. Fulfil your utmost human right, responsibility, obligation, and duty of creating crucial value to better humankind and advance the world.

INFINITY OF THINGS TO INVENT

The things to invent endlessly abound. Things to invent are infinite and unlimited. Everything good, great and valuable is possible when we execute the research rigorously enough to bring it into existence. Things to create, invent, or discover for the betterment of humankind and the advancement of the world are innumerable, boundless and inexhaustible. There are no limits to inventions that could and should be made. Whatever could help meet any of the countless human needs is a good target to be invented.

Every invention made creates room for more and better inventions to be made. Each impossibility made possible is evidence that other things currently perceived as impossible can also be made possible when we execute the necessary research. Every unsolvable problem made solvable proves that some other problems concluded to be unsolvable can also have perfect solutions created or invented for them if we thoroughly execute the necessary research for the purpose.

A workable solution invented for solving an unsolvable problem is a confirmation and testimony that functional solutions could and should be created for other unsolvable problems with the execution of the needful research. Indeed, every invention, creation, discovery, or innovation made is a strong motivating force for more discoveries, innovations, inventions, and creations to be made. We only need to meticulously execute the research it demands.

Whenever a person makes an invention that turns an impossibility into a possibility or an unsolvable problem into a solvable one with his research, several other people get fired up and propelled to aggressively embark on research thoroughly for the making of many other inventions. Every invention accomplished gives rise to the making of more inventions by the inventor and many others.

The making of many newer innovative products, devices, technologies, and solutions to persistent problems are inexhaustible opportunities that every person who truly wants can and should seize and capitalize upon to make far better ones. The number of existing inventions, products, services, and ideas at each time always creates opportunities for countless more astounding ones to be invented. Be one of the people to get the inventions done.

Do everything it takes to create and add value to the planet for the improvement of society, betterment of people, and advancement of the world. There can never be a shortage of the values to create, the products to invent, the services to provide, the things to discover, and the innovations to make for the advancement of the world. The inventions anyone who cares can and should embark on to make are always inexhaustible.

Every product invented increases the possibility for more products to be created. It is every person's ultimate human right and duty to research, invent, and create value for advancing humankind. Never ever fail to live up to your most essential responsibility of inventing something of great value to humanity.

The inventions we make dictate everything in life for humanity. Whatever we want can be invented when the required research is done. From every invention, no matter how remarkable, something much better can be invented with enough research.

Every invention leaves room for better inventions, just as each evolution is a basis for finer evolutions. Like every revolution on the planet that creates gaps for yet greater revolutions, every discovery serves as a springboard for more crucial discoveries to meet the needs of humankind better and skyrocket the world's advancement.

Every invention, regardless of its initial quality, has the potential to create immense value. Even from a not-so-good invention, great and marvellous things can be born. The key is to keep innovating, as excellent invention begets excellent inventions.

Unleash your curiosity and creativity, for they are the keys to unlocking new possibilities. The world is full of unexplored avenues waiting for your unique perspective and innovative ideas. Embrace these qualities, and you will discover that the potential for new and better things is truly endless.

Do something new. Create, invent, or produce new things. Things that best improve humanity and advance the world. Invent new procedures, new methods, and novel categories. Establish new systems, systems that dwarf all that have ever existed.

Strive for and achieve excellence with every research you embark upon. See the world with eyes different from all else. Make inventions that correct the bad and improve the good things you see.

With courage and arrogant confidence, do everything to make the undiscoverable become discoverable, unsolvable into solvable, uninventable into inventable, uncreatable become creatable, undoable become doable, and the impossibility into possibility for the best good of humanity.

Take extraordinary actions to get inventions made. Go

incomparably farther than any human has ever gone or could ever go to get your target invention made. At all times, know and operate on the unique principle that the uncommon wealth you crave lies far beyond where anybody else has reached or could go. Triumphantly overcome all the insuperable to get your dream invention actualized and enjoy the exceptional wealth and prosperity it brings.

Practice invention-making skills repeatedly to attain automation and superexcellence in each. It is the surest path to becoming a prolific inventor. We only master what we repeatedly do and become known for the value the mastery creates.

An unfailing key to making inventions is impeccable discipline and superb self-discipline. Splendidly discipline yourself to take all the risks, make all the sacrifices and invest all the resources, talent, time, and energy the actualization of your target invention depends on.

Living a risk-free life is a self-condemnation to making no invention and creating no value. A person must take all the risks that making an invention demands for him to move from where he is to attain the wealth he yearns for by exceptionally making the invention.

To create value with one's research and attain wealth, a person must take risks beyond what everyone else thinks is safe. He must pursue making inventions beyond what everybody else thinks is possible. He must chase and actualize dreams that everyone else thinks are impractical and unrealistic. He must fanatically commit to the materialization of the invention in view far beyond the level all else think is wise to do.

Next, digitalize the invention for every person in the world who cares to easily access and acquire it on the Internet with a few buttons on the computer or handset. Then, for many decades, generations and centuries to come, any individual or group of persons who want can access and acquire it at any moment from any part of the globe to gain from every bit of the product.

Making inventions is the ultimate human right, mission, obligation, duty, and responsibility. Nobody should wrongly think that humans will run out of what to invent if every person makes an invention. Regardless of how many inventions each person would make, human beings can always have what to invent. What to invent abounds and will always be. We always have what to invent. We can never run out of what to invent on Earth.

The things to invent in each human endeavour field are inexhaustible. For sextillion years to come, all the things to invent for the advancement of the world cannot be exhausted when every person fulfils his utmost human right, obligation, responsibility, and duty of making inventions.

Each invention accomplished creates opportunities for more inventions to be made. Therefore, the world will always have inventions to be made for its advancement.

The world will always have what to invent. Inventions to make will always be inexhaustible.

Please, quit the "status quo mentality". Go all out and create, invent, discover, and innovate products, ideas, services, solutions, technologies, devices, ill health interventions, and environmental remediations to best meet the deep needs of humankind. Create, innovate, and invent medical devices for the prevention, unparalleled diagnosis, treatment and cure of diseases. Make inventions that extremely improve healthcare, total well-being and quality of life.

In every area of life, the things to invent are innumerable and endless. Never an aspect of human existence that the products, devices, solutions, and services to invent are exhaustible.

In the sustainability of a safe, healthy and friendly planet, we cannot exhaust the things to invent even if a million inventions are made per hour in the next thousand centuries. Resource conservation, climate change, weather control, population crisis, air and water pollution, plus countless other environmental issues will endlessly call for more and many more inventions to be made.

Information Communication Technology Inventions

The trillions of inventions humans can make to exhaustively solve all information and communication technology-related problems will endlessly beat the wildest of all human imaginations. Many more inventions in information and communication technologies will continue to be needed. For instance, what we can invent to solve all Artificial Intelligence issues on this planet and, perhaps, other planets are countless. Even when every youth and adult invent hundreds each month for the next million years, there will still be some Artificial Intelligence issues begging for perfect solutions.

Data transmission and internet connectivity problems will continue to beg for a greater and much greater number of quality inventions to be made. Regardless of the volume of inventions, innovations, creations, and discoveries we make weekly in advancing information and communication technologies, some needs of humankind in that field will still be calling for better solutions. All of man's ICT needs cannot be exhaustively met. More cognate inventions in the field will always be needed.

Possible Inventions Needed in Education and Human Learning

What of education? Can humankind at any time finish making all possible inventions, innovations, creations, and discoveries to improve learning technologies, methods, and practices? The answer is as certain as a permanent 'No'. The potential for making inventions to enhance teaching-learning experiences can never run dry.

Man is a learning being. As long as human beings exist, there will always be inexhaustible room for more and much more inventions to be made for the perfection of human learning.

There will never be a shortage of possible educational inventions for learning human languages and other living creatures' socialization processes. Lower animals produce sounds and make signs that are meaningful to them. And human beings have got to learn all that to provide better care for their existence.

The century, human beings would have exhausted all possible educational inventions to improve group and personalized learning is never in sight and is not likely to be. Countless educational techniques, devices, and technologies for extreme optimization of memory and learning in an absolutely unforgettable manner will continue to beg for more inventions to be made.

Needed Sustainable Energy Inventions

The demand for sustainable energy will continue to elicit countless inventions to be made. At no time will the possibilities for creation, invention, and discovery of exceptionally better sources of and solutions for more sustainable energy be exhausted, regardless of the number of relevant inventions made weekly.

Renewable and non-renewable energy will continue to demand more inventions over time. The need for far better energy creation, conservation, storage, efficacy, and efficiency will never have all the imaginable inventions made. There is no energy source (biofuels and biomass, nuclear, solar, geothermal, hydropower, thermal, fossil, natural gas, coal, oil, petrol, wind, tidal, etc.) that we can completely exhaust the creation of new possibilities on. Why not visualize something in this direction and work committedly on it until an awesome completion of the invention?

Take batteries alone, and you will see that inexhaustible inventions, innovations and discoveries must be made. All the imaginable devices for converting chemical energy into electric energy cannot be exhaustively created. From Flow batteries such as zinc-polyiodide, organic aqueous, vanadium-redox, and so on to Lithium batteries such as lithium-metal, lithium-sulfur, lithium-ion, and countless others begging to be invented to Zinc batteries that include zinc-manganese oxide, zinc-air, and so on, to Lead batteries like lead-acid, be it absorbent glass mat, valve regulated lead-acid, deep cycle lead-acid, high-rate lead-acid, Ortsfest PanZerplatte Verschlossen [OPzV tubular], GEL sealed lead-acid, etc.). Plus, a battery type that will be named after you are all new possibilities that are inexhaustive.

Transportation of people, goods and services

Think of the different places you have been to and the innumerable places a person could desire to reach with countless means of transportation that could be invented. Could the possible inventions to be made in transporting humans, goods and services be exhausted? No, certainly. Inexhaustible means of better mobility and transportation will endlessly beg for many more inventions to be made.

Inventions in the area of perfectly safe and automated electronic transportation alone will endlessly call for much more rigorous research and thorough inventions to be made. Much better, safer and faster means of transportation by land, water and air will eternally demand new possibilities. We cannot lack what to invent any day for automation of mobility and transportation of people, goods and services.

Growth and use of plants

Imagine the world of plants. There are countless plants in the world and humans have not and will not completely finish discovering the best ways of growing any plant. Furthermore, never a plant that man has completed the invention of all the uses to which it could be put. Every plant is so richly endowed that with much more rigorous research works, inventions and discoveries could be made on exceedingly better ways of usefulness to which the plant could be put.

Inventions will endlessly be done on all the best uses of each plant and combinations of the uses of two, three, four, five, and more plants for the betterment of human lives. More food and overwhelmingly more drugs could be invented or discovered about everyone or more plants on Earth.

Natural resources inventions

More natural resources are begging to be discovered, mined and refined. Even the natural resources currently explored and exploited, there is none that incomparably much more refinement and use cannot be discovered or invented. God has so much endowed the world with great things that continuously need many more inventions and discoveries to be made out of. With much more rigorous research, each metal could be used far better. Something overwhelmingly finer could be invented from a combination of some metals.

Inventions in the entertainment industry

Creativity will timelessly be needed in the entertainment industry. The need to invent new possibilities will continue to abound in the entertainment industry.

Better ways of capturing, disseminating, and preserving news, theatrical video, motion and still picture, and audio recording and display will keep calling for more creative inventions. Music, dancing, film, sports, video games, television, radio, national dailies, magazines, journals, books, tourism, and special events cannot exhaust all the new possibilities.

How to provide and deliver the very best relaxation to different kinds of audiences worldwide all at once will continue to beg for more inventions to be made. There is no amount of service that the service industry and entertainment industry will render to permanently satisfy all the needs of humankind in that regard. Therefore, more inexhaustive creativity, discoveries, innovations, and inventions will have to be done over the years, decades, and centuries to come.

Never at any point will a person lack what to research on and create greater value for humankind than all that currently exists. Therefore, we individually and collaboratively have endless possibilities for making inventions to better meet the great needs of humankind.

Choose a great challenge facing humankind. Next, embark on fulfilling your absolute human right, responsibility, obligation and duty of inventing an enduring solution for the best improvement of humankind, betterment of society and advancement of the world.

CHAPTER 8

INVEST EXTRAORDINARILY TO INVENT

The making of an invention demands extraordinary investment in it. Extraordinary investment of talent, treasure, time, and action is mandatory for a person to invent something of value. To invent anything of great value worldwide, one must invest extraordinarily in the project.

1. Making inventions demands extraordinary investment.
2. Extra 50 per cent investment in, invents anything.
3. Complete self-commitment to making inventions.
4. Seek and attain new possibilities at all costs.

MAKING INVENTIONS DEMANDS EXTRAORDINARY INVESTMENT

The target invention gets made with extraordinary exertion of action in the invention-making journey. Extraordinary exertion of action in maximally investing one's resources, talent, time, and treasure far beyond what all others have done is a necessary condition for making an invention.

It is only when a person exerts incomparably more action and invests exceptionally greater than the amount of time, treasure, talent, and action in the right mix that the target invention is actualized. The investment of the average of the time, resources, and action that others have exerted towards making an invention is absolutely inadequate to make the target invention. Even exerting the greatest amount of talent, treasure, time, and action in the same combination, mix, or proportion that someone else has invested in without getting the invention done can never make the target invention.

Emphatically, to invent, create, discover, or innovate a product, service, device, or solution, a person must invest extraordinarily more talent, treasure, time, and action than all that has been done by others who fell short of achieving the target. Only exceedingly greater potential, resources, time, and action than ever exerted by others can make the desired invention. Invest extraordinarily more than all that has been done unsuccessfully for the making of an invention to get the invention made.

It is the extra far beyond the very best that has been done in the past that gets an invention made. Merely exerting the highest level of effort that someone else has put in that did not get the invention made cannot enable you to accomplish the invention. An exceptionally greater investment of action, time, talent, and treasure than has been done before is mandatory for making the invention.

A person has to develop his research execution skills (reading, writing, experimenting, analyzing data, and disseminating information) to the autopilot level and invest everything about him in the invention-making journey to achieve the target invention, creation, discovery, or innovation.

Never has, and never will, an average vision and average investment of action, resources, time, and potential result in the creation of something of immense value for the momentous improvement of human lives and advancement of the world. Average investment in, standard exertion of action, and normal sacrifice of time, resources, and talent have not and will never invent or create a thing of indelibly tremendous value for humanity.

To create or invent a product of remarkable value for the advancement of the world and the improvement of humankind, an extraordinarily greater than the average investment of action, talent, and resources in the target product is mandatory. It is an extreme investment of everything in making the target invention, substantially more than all others have done, that leads to attaining the goal.

Every inventor invested extraordinarily greater than what an average person does in the invention-making journey to accomplish the feat that has set him apart from all other human beings. The normal investment of action, which most people do, never results in the creation of any life-changing value in the world.

Moderately investing everything about a person in the invention-making journey never gets anything invented. The average is the midway between two extremes. The two extremes are:

(1) Making a particular invention.
(2) Not making the invention.

Making the invention demands completely investing a person's talent, time, resources, skills, energy, and action. Investing half of a person's talent, time, resources, and action in pursuing an invention never makes the invention. The midway between inventing and not invention a product does not and can never achieve the target invention.

An incomplete invention-making journey never results in making the target invention. Any person who embarks on an invention-making journey and quits midway never achieves the ultimate goal of the research. He who quits midway never gets the invention made.

Average exertion of action towards making an invention never accomplishes the invention in view. Investing half of a person's time, talent, treasure, and action or effort in an invention-making journey does not accomplish the target invention. A person's average investment of resources, talent, action, and time is never an option for making an invention.

Only the exertion of extraordinary action and exceptional investment of a person's time, resources, and talent in pursuing the target invention results in the actualization of the invention. Every renowned inventor extraordinarily invested his talent, resources, potential, time, and action in the successful invention-making journeys he embarked on.

Take after prolific inventors. Extraordinarily invest everything in and about you in the pursuit of invention-making until its exceptional accomplishment.

Please excellently complete each invention-making journey you embark upon. No giving up on your invention-making journey until exceptional accomplishment.

Keep going, nonstop, on your journey for value creation until it is done excellently well. Never ever quit midway in any of your journeys to make inventions. Do not hold back anything that successfully making the invention demands investing in the pursuit.

EXTRA 50 PER CENT INVESTMENT IN, INVENTS ANYTHING

With the right actions taken to the right degree over time, regardless of the challenging circumstances, the envisioned invention gets made. Whenever a person invests an extra 50 per cent of all the necessary resources and actions to make an invention, it gets accomplished over time.

Every invention-making goal persistently pursued with total

commitment and complete dedication with the right focused actions to the right magnitude materializes. Every target invention gets accomplished over time when the right actions are engaged in 50 per cent more than required.

Virtually every one of the countless inventions made on earth was achieved by exerting the requisite actions far beyond the right amounts to get it done. Committed right actions 50 per cent more than the necessary amounts result in the astonishing completion of any invention-making journey. Fifty per cent extra investment of everything necessary for making an invention gets the target invention actualized exceptionally well.

Whenever you set out to invent something of great value to humankind, do not only exert the magnitude of sacrifices, investments, energy, commitment, passion, and actions necessary for attaining it. Rather, exert an additional 50 per cent enormity of sacrifices, investments, energy, commitment, actions, enthusiasm, and persistence necessary for making the invention. Then, the goal of the research, the target invention, gets fantastically made.

Take an extra 50 per cent of the required risks and exhibit an additional 50 per cent of courage demanded to make the target invention. With the 50 per cent extra sacrifices for and investments in making any invention, it gets exceptionally completed. The role of 50 per cent more right efforts than necessary for creating any greatly useful thing to humanity cannot be over-emphasized in getting the invention astonishingly accomplished.

COMPLETE SELF-COMMITMENT TO MAKING INVENTIONS

Total self-commitment to invention-making as an absolutely necessary obligation that must be fulfilled is the only way out. The attainment of genuine greatness, glory, and wealth depends on complete commitment and accomplishment of the invention-making journey.

We must commit ourselves completely individually to making an invention of great value at each time. We must individually do everything, no matter what, to excellently accomplish the target invention of our research each time.

At each period, a person should be working committedly on creating value. The value should mean substantially more than everything else to him.

Do not at any moment surround yourself with losers. People who have not created any value and are not dedicatedly working on inventing something of great value should not be those you closely associate with.

Identify an inexistent thing of immense worth to humankind and committedly work on bringing it into existence as your sole obligation until the actualization of the invention. When a person refuses to work committedly on making the invention that matters the most to him, he spends his lifetime trying to fulfil everyone else's purpose. He ends up inventing nothing, creating nothing, and making no useful innovation. He disappoints himself and everyone else.

Be committed completely to the making of inventions. The amazing making of any invention demands complete commitment. Totally commit yourself and everything about you to the making of a valuable invention each time you embark on an invention-making journey. Make a complete commitment of yourself and all you have and are to the making of a crucially useful invention every moment of your life. Keep working from strength to greater strength on its completion until the invention gets fantastically made.

Commit all your energy, talent, resources, and everything you have to actualize your invention vision. Immensely, consistently and persistently work each day on attaining your target invention as though your entire life depends upon it alone. Invest every hour of each day working with gusto on the exceptional completion of your target invention until it is marvellously achieved.

Aggressively exert enormous daily action on making your target invention as if your life and the lives of countless others depend upon the exquisite actualization of that invention. Certainly, your life and the lives of countless other people depend upon the excellent making of the invention your research aims at. Keep taking monumental action daily for the exceptional completion of the invention until it is wonderfully done.

Passionately exert tremendous daily action on the thunderstruck accomplishment of your target invention. It is your utmost human right and greatest responsibility, duty, and obligation to better people and society and advance the world with the value you create.

By refusing to dedicatedly work on making an essential invention, an individual jeopardizes his life and the future of countless others. Only when a person singleheartedly works and fulfils his ultimate duty of making an invention of great worth to humankind that he can secure a better future for himself and all others.

Making inventions is an absolute necessity for actualizing a better future for all. The pursuance of invention-making has got to be approached as an absolute must until the fantastic accomplishment of the target invention.

SEEK AND ATTAIN NEW POSSIBILITIES AT ALL COSTS

It is with the making of investment in a research journey at all costs that an invention gets made. Research is the empirical quest for new possibilities. The new possibility, such as an invention, creation, innovation, or discovery of great value, is only achieved when the researcher invests in it at all costs.

An inventor is an individual who typically engages in the empirical search for brand-new possibilities to improve people, better society, and advance the world. The inventor invests everything in and about him in pursuing the new possibilities before actualization.

To make remarkable success in the quest for novel possibilities, inventors never look at the world as *it is*. Rather, inventors characteristically visualize the world as *it should be* and invest everything into actualizing or materializing their blissful vision. Develop a mindset backed completely by seeking and attaining new possibilities at all costs for the best good of humanity and the world's advancement.

Inventors always think of new possibilities and unreservedly invest in and intensely exert persistent actions to bring the new possibilities into existence or reality. Inventors experiment with different things and ways to actualize their wonderful vision of the world. Suppose one method or way does not work. In that case, they persistently try something else until the new possibility is actualized.

Thinking, tinkering and acting outside the box always sets inventors apart from the rest of humankind. They think of and act on the principle that at each time and in each space, more than two possibilities exist. Plus, more than two possibilities can be combined and rejigged to bring something absolutely better into existence.

Inventors make things that have never worked out in the past to work almost perfectly from the present onwards. They make all the necessary sacrifices and invest far more than necessary in pursuing the research goal until the envisioned inventions are actualized.

Recreation of things that already exist in some known forms into completely different and more valuable forms that were and are unknown is what inventors do. Please, do everything required, make all the sacrifices for, and unreservedly make every one of the investments in getting a thing of crucial value invented to become an inventor.

Inventors often conceptualize and act dedicatedly to successfully improving upon things that exist. They draw inspiration from what exists to build new and better things, regardless of cost. They know that when done, the invention, exquisitely made, will certainly attract unimaginable wealth from across the world to them.

Kindly do all it takes and become an inventor. Crave and create new possibilities to better human lives and advance the world. Then, the invention made shall turn into an endless stream of wealth for you.

We have to be constructively creative and visionary in selflessly investing in making inventions to become renowned inventors. We must see flaws in existing things and intensely experiment on successfully creating products, solutions, services, or systems that overcome the observed flaws. Learn to combine many existing methodologies and improve significantly upon them in making new and incomparably better products.

Inventors never live comfortably with the results of other people's thinking, opinions or products. Inventors always seek newer and absolutely better ways of getting things done and problems solved. Next, they unreservedly embark on the invention-making journey until the awesome actualization of the blissful vision.

Inventors invest completely in the actualization of their invention epiphanies and intuitions. Please, passionately take after or learn from and emulate the well-known inventors to eventually become a great inventor.

Make the research you do a big, lucrative global business. Be skillfully equipped with all it takes to maximally function in the global digital knowledge economy, as emphatically presented in this book, **Research, Invent, and Create Wealth**, in addition to the previous book, **Research: Make Impossibility Possible**, by the current author.

The intensely competitive global digital and knowledge economy demands researchers to place unreserved emphasis on making inventions, discoveries, creations, and innovations, at all costs, for advancing Science, Technology, Engineering, Mathematics, Medicine, and Arts. Researchers must be totally committed to or concerned with advancing both technologies and humanities. For instance, only an individual with skilful reading, writing, and quantitative abilities can contribute much to and meaningfully benefit from a technologies-driven world.

This book champions the making of worthwhile innovations, discoveries, creations, and inventions by nurturing the users' minds to imagine a world radically different from the current one in which we live. It retrospectively demonstrates that the world has been different, could be different, and will surely be very much different again for the better if we invest completely in the making of inventions now for that purpose.

It maximally motivates creativity and productivity in the users. The book lavishly delivers the distinctive obligation of nurturing and fulfilling the deeply seated human desire to best understand ourselves and the world around us via the creation of relevant values for the accomplishment of our cravings. We can individually and collectively achieve this special obligation by executing the needed research.

CHAPTER 9

EXTREMELY CRAVE MAKING INVENTIONS

A man with a big enough reason to invent something can sacrifice every pleasure and bear every pain in researching to make the target invention. There must be a strong purpose before an invention can be made.
1. Develop extreme craving for making inventions.
2. Actualize your craving to be a renowned inventor.
3. Delayed gratification for making inventions.

DEVELOP EXTREME CRAVING FOR MAKING INVENTIONS

Develop a great craving for making inventions. Extremely crave making inventions. Extreme craving for making an invention elicits exertion of all the necessary routine actions that culminate in the making of the invention.

Cultivate a great need, an irrepressible desire, backed with relentless unreserved action, for inventing something of value. Such a great craving propels the person to exert every action, no matter what it takes, to make the invention the research targets.

Be emotionally, mentally, physically and spiritually engaged in doing everything making the target invention demands. Strong emotion for invention-making drives behaviour. By 'behaviour', I mean the great action mentally and physically exerted to accomplish the target invention – the attainment of the ultimate research goal.

When every needful action is meticulously exerted in a research journey, God, the creator and sustainer of the universe, blesses the researcher with an outstanding completion of the target invention, innovation, creation, or discovery. Nature opens its doors for the actualization of the invention only after flawlessly exerting actions much more than all other humans have done towards making the invention in question.

A research work investigates a problem to get it solved. When everything the research requires is done exceptionally well, the study's goal gets achieved, and the target problem gets solved.

Feel constructively dissatisfied with the status quo. Visualize the suffering that characterizes humankind under the status quo. Let the suffering heighten your constructive discontentment with the status quo. Let the suffering drive your emotions, behaviour and entire being to crave a change of the status quo desperately.

Have the strongest craving for a change. Then, take all the actions with your research to materialize the desperately desired change. Research and make the invention upon which actualization of the craved change in the status quo depends. Let the change put an end to the sufferings that characterized the status quo before embarking on the research.

It is the irrepressible desire to change the status quo for the best that powers a person to take all the actions in executing research. The research must be such that when done, it invents a product, solution, device, service, process, idea, technology, or new knowledge that brings the desired change in the status quo into being.

Wholeheartedly wanting more, craving the best, and desiring the finest galvanized renowned researchers to seek inventions for improvement, advancement and development of the world via their research. When a person fiercely desires or craves a change for the best, he embarks on the research that will create, invent, or discover the things that will surely bring about the desired change.

Such research demands making unreserved sacrifices. It demands investing in everything else. Kindly do whatever it takes to complete the research that actualizes the desperately desired change awesomely.

Reward in the invention-making journey only comes after a committed investment of high-quality time, energy, talent, effort, resources, and treasure in executing the necessary research. We must invest all the needful action, energy, time, and resources in the exertion of our talents costly enough to bring about the incredibly desired reward – the attainment of the inventions for a much better world.

Attainment of the ultimate research goal is a reward that best satisfies the researcher's craving. The accomplishment of the ultimate research goal satisfies, gratifies, and fulfils the researcher's quest, improves people's lives, progresses society and advances the world. Please, desperately crave something that is not yet in existence. Then, embark on research thoroughly to create, invent or discover the object of the craving.

At each time, what we badly need outpaces what we have. To meet our needs, we must research and make the necessary inventions that satisfy the most serious needs.

Badly need creations, inventions, discoveries and innovations for the world's advancement. For a person to badly need something does not mean the person has too little. Rather, it means the person wants a lot more and is committed to taking all the necessary actions to meet the need. We become satisfied, fulfilled, prosperous, and happy when we research and invent to meet our needs.

A person should do everything to achieve the invention his research targets. He feels most hurt when he fails to accomplish the target invention. If you do not want to be hurt, invest everything to actualize the invention your research aims at unfailingly.

Cultivate the greatest desire for what you want to invent with your research. Feel exceptionally motivated, galvanized and amped-up to exert every action toward, make every sacrifice for, and invest everything in the research for making the target invention.

Let the invention made with your research cause you to feel extreme satisfaction. Plus, let the satisfaction motivate you to committedly embark on significantly more challenging research to make more exceedingly valuable inventions.

And let the cycle of great craving for a nonexistent thing, executing the research for marvellously making the invention that satisfies the craving, continue till you become a globally famed prolific inventor. Let the cycle of great craving for the nonexistent wonderful thing, successfully embarking on research for inventing the product, device, solution, object, idea, technology, or service that gratifies the craving continue until you become renowned the world over as the inventor whose research works have made the greatest positive impact on the world's advancement.

The most desirable thing to do on the planet is the making of inventions. You must crave making inventions. A person must ultimately seek and attain the making of inventions. He must love making inventions of great value.

High and above everything else, a person must desperately desire to make inventions. A person who craves, extremely wants, gravely likes, and critically desires to make inventions far and above everything else is motivated intrinsically and extrinsically to committedly engage in the exertion of extraordinary actions for making the inventions. Please research and invent something of inestimable value to humankind and become one of the world's renowned inventors.

Such an individual is best motivated to meticulously engage in research execution or funding for unfailingly making inventions of immense worth. Extremely desiring and craving the making of inventions drive a person to rigorously embark on research execution or funding from time to time to achieve the target invention in each research journey exquisitely.

The successful completion of each invention-making journey gives the person an unimaginable pleasure and satisfaction that, in turn, propels him to sustain research execution, one after another, to make more and much more valuable inventions. Such persons create or bring about the actualization of the blissful tomorrow of our dreams. Dedicate yourself to becoming one of them.

Stun the world with your research findings, inventions, innovations, creations, and discoveries. Research and make an essential invention or discovery that gets the world stunned.

ACTUALIZE YOUR CRAVING TO BE A RENOWNED INVENTOR

Crave to become an inventor of great value. Let becoming an inventor of something of great value be the ideal person you wholeheartedly want to be. Plus, do everything necessary for you to become that ideal person.

Focus on who exactly you wish to become by changing your habits to be in accordance with your ideal person, a renowned inventor. Change your habits to become the kind of person you seriously want to be. When you become the type of person you eagerly wish to be, you can most easily work to achieve every one of your goals of making indelible inventions, discoveries and innovations.

Change your beliefs about yourself. Daily take small actions that shall, over time, change your habits, upgrade your substance, expand your knowledge, improve your skills, and optimize your self-development for the best actualization of your peak potential. Continuously perform the small task units that marginally accelerate you closer daily toward completing your invention-making journey.

Change what you think, say and do to become your ideal person, the renowned inventor you crave to become. The sure way to practically change who you are is by changing what you think, say and do.

Nothing changes, and nothing is going to change when a person keeps thinking, declaring and doing things the exact old ways they have typically been done. No invention is made by thinking, declaring, and doing things the exact old ways a person has been doing. When a person practices the same old habits that have prevented him from rigorously researching and making inventions of tremendous value, he cannot become the prolific inventor he craves to become.

Think differently, confess more positively and exert considerably more different actions daily to steadily move closer to the outstanding accomplishment of the inventions your research targets. Achieve each of the target inventions of your research journeys excellently well to become the prominent inventor you crave to be.

To become a prolific inventor, a person must be a chronic researcher. A cue triggers a craving. Craving motivates response. The response produces reward over time. Reward satisfies the individual's carving.

To become a veteran inventor, a person must habituate invention-making. He must repeatedly complete the invention-making cycle of:
1. Cue producing craving.
2. Craving triggering and propelling response.
3. Response causing reward (the accomplishment of the target invention).
4. The reward satisfying or fulfilling the craving.
5. Fulfilment of the craving extinct the cue.

The importance of completing the invention-making cycle cannot be overemphasized enough. I have to restate it to drive the point home.
1. Cue elicits or gives rise to craving.
2. In turn, craving propels and motivates routine exertion of invention-making actions.
3. The exerted invention-making actions attract the expected reward - the target invention gets made awesomely.
4. The exquisite making of the target invention satisfies the craving.
5. Satisfaction of the craving brings the cue to extinction.

When a person typically and effortlessly starts, continues, and completes the invention-making cycle from time to time, making valuable inventions becomes his habit.

Habitual completion of the invention-making cycle is indispensable for a person to become a globally renowned inventor. That is, he visualizes cues that drive craving. The craving leads to his routinely performing the needful invention-making actions. Then, the meticulous actions materialize the ultimate research goal, making the target invention, discovery, innovation, and creation. The making of the creation, discovery, innovation, or invention satisfies the craving. With the satisfaction of the craving, the cue reaches extinction.

When the invention-making cycle is completed over and over, repeatedly and repeatedly over time effortlessly, then the person has made the making of invention his habit. Having and regularly demonstrating the invention-making habit gets the person to become a profound and prolific inventor.

Forming and regularly demonstrating the invention-making habit is necessary for a person to become a famous and prolific inventor. A person can only become a prominent researcher and prolific inventor when he has fully developed the habit of making inventions, discoveries, innovations, and creations.

Please, crave invention making. Habitually execute research successfully. Habit is the automation of the processes of cue leading to craving, which leads to a routine response that produces the reward. The reward is attaining one's craving, which is the target invention.

Far above everything else, please crave being an eminent inventor. Make all the sacrifices and invest every talent, time and treasure in researching, making inventions and fulfilling that craving. When a person who craves to become a renowned inventor does all the research, makes all the inventions, discoveries, creations, and innovations and eventually becomes a prolific inventor, he has been rewarded.

Dedicatedly experiment, write, read, analyze data and disseminate the information regularly. The more you intensely repeat these behaviours, the better you become at them.

The better you become at experimenting, writing, reading, analyzing data and reporting findings, the greater you embody the unique identity of an inventor. The greater your embodiment of a unique inventor identity, the more valuable inventions you make.

When you make many enough indispensably useful inventions, you become the prolific inventor you have ever craved. With that, you achieve the peak of your creative potential.

Originate Invention-Making Association. Spread it until hundreds and thousands of millions on the planet across all fields embrace making inventions as the ultimate life goal.

Let countless people habituate invention-making skills. Let higher institutions of learning award degrees based solely on inventions, discoveries, creations, and innovations made.

Provide methods for attacking and changing uninventive habits as emphasized in this book and the other books by the current author. Let the methods effectively change people's uninventive habits into invention-making habits.

Get and let the Kpolovie's books become a giant machine for changing invention habit loops from negative to positive. With the the books demonstrate why and how even the most obstinate, uninventive habits get changed into profound invention-making habits.

DELAYED GRATIFICATION FOR MAKING INVENTIONS

Delayed gratification is the route to ultimate research goal attainment. The exceptional making of the target invention largely depends on the extent to which the researcher delayed gratification while dedicatedly investing everything in accomplishing the research.

Making an essential invention with research requires the investigator to ignore immediate reward in favour of delayed gratification. To make an invention with research, a person must train himself to reject and avoid immediate reward in favour of a delayed reward. Invention-making demands delayed gratification. We get to train ourselves to overlook, disregard or reject immediate rewards in favour of the target invention of our research.

Making an invention is like a marathon in which only one person, the one who delayed gratification most, shall be the winner of the prime trophy. Very many people embark on research to make a particular invention of the greatest value. Only the person who most delays immediate gratification gets the invention made. It is the individual who best sacrifices everything else for and invests every other thing in making the invention that gets it made.

What else can one liken the quest for making an invention to? Yes, it is like hunting for a life lion in the jungle. The successful hunting of a life lion in the jungle requires not catching smaller or less valuable animals which may come the way of the hunter. A hunter who attempts to catch every other animal he sees in the jungle is bound to fail in catching a life lion in the jungle.

Only the hunter who concentrates solely on capturing a life lion can easily succeed in the mission. The animals other than the lion that the hunter sees are like immediate gratifications. A hunter must neglect all the immediate gratifications and concentrate on capturing a life lion before he can achieve the challenging hunting goal.

In virtually every sphere of endeavour, remarkable success demands ignoring immediate reward. Becoming an outstanding researcher and a renowned inventor demands delaying immediate gratification. We must sacrifice the impulse for immediate gratification in order to successfully research, invent, and turn possible impossibilities and solvable, unsolvable problems.

Man is a reward-centered animal. A top reward-centered animal is man. The human being is naturally motivated by reward. The reward system that propels our behaviour is such that engagement in what is rewarded gets repeated. On the contrary, engagement in what is punished is avoided.

We are more likely to repeat an experience with a satisfying ending. Conversely, man is more likely to avoid an experience with a painful ending. While an experience with a satisfying ending is a positive reinforcement, reward, or motivation, an experience with a painful ending is a negative reinforcement, reward, or avoidance conditioning.

Very often, both positive and negative experiences are what motivate us to do some things and avoid some other things. Positive rewards (introduction of what increases satisfaction or introduction of what reduces pain) and negative reinforcement (removal of something satisfying or introduction of what increases pain) are often at work in motivating us to engage in certain actions or behaviours.

Actions that bring rewards, satisfaction, fulfilment, and pleasure are repeated. Conversely, actions that do not bring rewards, satisfaction, pleasure and fulfilment are avoided.

For renowned researchers and prolific inventors, engagement in actions that lead to the making of inventions is repeated. In contrast, engagement in actions that do not lead to invention is avoided.

The best rewards that our actions could bring are inventions. When we committedly engage in rigorous research, we get or attract the greatest of rewards – the making of inventions and discoveries.

Everything we use was created, invented, discovered, or innovated with research. A person should, therefore, embark on research to successfully create or make inventions, discoveries and innovations that best meet the needs of humankind and advance the world.

Almost everything we use is a product of research done

mainly by others. Why should a person not thoroughly embark on research and make some inventions and discoveries that others will use? There cannot be any justification for a person to refuse to research or fund research to make inventions.

As the making of an invention is the greatest of rewards that a person can ever get, why should a person not take all the actions research demands to make an invention? There is no reason good enough for me, you or anyone not to make an invention of great worth. The greatest reward we can ever have should be the strongest motivating factor for our actions, engagements and behaviour.

The making of an invention as the reward for research execution is usually a delayed reward as opposed to an immediate reward. Making an invention does not happen just immediately after an action is exerted.

It takes the exertion of many extreme actions for a crucial invention to be made. It does not take a single action for research to be exceptionally completed. It takes the greatest efforts and time in terms of weeks, months or years of dedicated hard work to complete a research project that invents, creates, or discovers something of astronomical worth.

This accounts largely for why people who are instinctively focused on just the present or very near future or immediate reward do not successfully embark on or invest in research. Consequently, people who seek only immediate rewards do not make any inventions.

Those whose predominant concern is doing only things that give immediate or near immediate gratification, individuals concerned primarily with getting immediate rewards, do not make inventions. Far from researching, the people who predominantly seek immediate rewards like what to eat now, where and when to sleep today, using invented things, and how to avoid short-term or immediate pains do not make inventions.

They only think of and do what gives immediate satisfaction, pleasure or fulfilment. They never do what will improve tomorrow or the future for humankind. Please, always be different from the people whose concern is immediate gratification.

Conversely, individuals who research, create, invent, and discover are typically focused on thinking and doing things that will bring about an improved and advanced tomorrow. They delay rewards to ensure a much better tomorrow. They are concerned with the ultimate outcome of their actions that usher in a better future for everyone. They delay gratification. They exert actions that will, over time, cause incomparably better and lasting consequences.

The people who make inventions envision the future and make and implement plans for engendering a much better tomorrow for humankind. They seek and take actions that have long-term benefits for the majority. They invest heavily in research now and make sacrifices in the present to provide a far better future for people, society and the world with their research.

Rather than taking actions that will give them immediate pleasure, inventors take actions that will meet long-term goals, and bring lasting satisfaction, pleasure, fulfilment and accomplishment for all in the future. They engage in actions that are immediately not rewarding but will surely yield lasting benefits for the majority in the future. They embark on and complete research that invents what tomorrow's betterment depends on.

Please, do today the research that will invent the things upon which people's improved lives, a far better society and an advanced world tomorrow depend. With our research today, let us invent the blissful tomorrow we crave.

CHAPTER 10

RESEARCH SUSTAINS THE WORLD'S ADVANCEMENT

Sustainability and advancement of the world depend predominantly on the making of inventions via research. Research makes possible impossibilities. Research makes the unsolvable solvable. Research is the creation, invention, and discovery of valuable products, solutions, services, devices, knowledge, technologies, or ideas for the sustainability and advancement of the world.

1. Invent to sustain the world's advancement.
2. Invention-making action alone produces results.
3. Work unreservedly happily to invent.
4. Sing the victory song for successfully inventing.

INVENT TO SUSTAIN THE WORLD'S ADVANCEMENT

Research is the empirical quest for the creation of new solutions, hitherto inexistent solutions. It is a never-ending hunt to create or make inventions and discoveries that deliver better results more easily at a lower cost to the audience or beneficiaries. Making an invention or discovery that practically delivers perfect or extremely valuable results efficiently at a low cost to the users is what research does. This way, research advances the world.

Aptly, research is the empirical quest for new possibilities. It is with research alone that new possibilities are created. Great things that were conclusively considered to be impossible because they are too good are made to become possible things with research. Research is done to turn impossibly good things into possibly great things for the best improvement of humankind and the advancement of the world.

For everything, product or service in existence, an incomparably better thing, product, solution or service can be invented and massively made available to humankind at a much cheaper cost. We only need to execute enough rigorous research for the purpose. Meticulously research, thoroughly execute research and create, invent, or discover exceedingly useful solutions, products, technologies, or services for the betterment of people's lives and the advancement of the world.

It is only with research that what is so good and currently concluded to be impossible to exist is brought into existence. Research invents or discovers valuable technologies, things, products, devices, services, procedures, or solutions that are impossible by the status quo.

With research, exceptionally valuable products, solutions and services that are impossible by the current convention or status quo are turned to become possible. The world advances or improves remarkably only with the inventions, creations, discoveries, and innovations made with research.

Without inventions and discoveries via research, the world can only be retarded, impeded, retrograded, and forced into extinction. The world's sustainability and advancement demand more and much more research. The sustainability and advancement of the world depend solely on the products of research.

We owe the world researching and making substantially

more valuable inventions and discoveries than all that exists for the advancement of the world and the improvement of people's lives. Making incredible inventions and creating excellent value is a debt we must pay. We must research and make inventions that will sustain humankind and rapidly advance the world.

We must passionately engage in executing research. We have got to embark on a never-ending quest to deliver much better results more efficiently at a more affordable cost to people. Research is the only way to make inventions and discoveries that pay the great debt we owe the future by advancing the world today. We must embark on research to bring about the far better tomorrow we crave.

Dedicate much more time, effort and energy to your self-development for successful research execution. Too much time, energy and resources are spent on things that do not develop our research and invention-making skills. We get to adjust or change our way of life for better self-development and optimization of our research and invention-making skills.

Often, people spend hours or long minutes sitting through uninspiring movies. More often than not, we waste very many minutes checking our phones, mindlessly staring at the screen. Daily, people kill minutes attending to notifications on their computers or handsets. Valuable hours are frequently lost to group chats, games, and unnecessary social media.

Unsolicited emails we receive take some minutes of our time. The average person spends as much as three hours daily on social media. Three hours a day is over 1,090 hours in one year. All such hours can and should be gained for self-development and successful research execution.

The time and energy that social media, phone screens and irrelevant emails sap from us weekly could best be invested in developing ourselves for research execution and invention-making.

The advancement of the world, improvement of our future and realization of the tomorrow we crave depends on the inventions, creations and discoveries we make with our research today.

It is only right to stop wasting time and energy on things that do not lead to invention-making. Committedly devote all such time, effort, energy and resources to self-development for research execution, making inventions, and creating value.

Delete the time-wasting games and the unnecessary social media apps on the phone. Quit or mute some group chats. Unsubscribe from time-sapping emails. Put the phone on silent to reduce distractions whenever you are meant to be completely dedicated to research execution task performance.

Put off the phone for some hours and use the time and energy for self-development and research execution. Use all the gained time for reading, writing, experimenting, analyzing and practicing the various research execution skills.

Invest more and much more of your time, effort, talent and wealth in the quest for making inventions and discoveries that pay the debt we owe the future. Advance the world by improving humankind's lives with your research products.

Make a commitment and determination, a firm resolution backed with action from today for daily taking actions that consistently move you closer to attaining your ultimate research goal. Such determination is a resolute or firm commitment a person makes to unfailingly actualize his ultimate research goal by working enthusiastically daily for the purpose, irrespective of the odds and challenges he might experience.

Such firm determination could also be referred to as a commitment device. It is a strategy for getting more done consistently toward accomplishing a person's invention-making journey.

A commitment device is an unbreakable bond an individual

places on himself in the present to restrict his future choices and actions that might derail him from attaining the research goal. It is an unwavering determination, backed with ceaseless action from the day it is made until the successful completion of the invention his research aims at.

For example, a commitment device could be that:
- 'I must consistently self-develop my research execution skills daily.'
- 'Daily, I get to accomplish some units or aspects of the research that markedly move me closer toward the outstanding making of my crucially valuable invention, discovery, innovation, or creation.'

Then, the person strictly adheres to the resolution until the outstanding completion of the target invention. After achieving the invention, he embarks on another research work to make a much more essential invention. He continues the research execution circle to make inventions upon inventions successfully. Then, he actualizes himself as the prolific inventor he was created to be.

We must develop ourselves excellently for research accomplishment, invention making, and turning impossibilities into possibilities and unsolvable problems into solvable ones. We must overwhelmingly practice, read, write, experiment, analyze data, and disseminate results. Attainment of peak performance level via intense practice of reading, writing, experimenting, data analysis and results dissemination skills is indispensable for accomplishing invention-making journeys.

When we exceptionally develop ourselves for research, exquisitely completing an invention-making journey surely becomes an easy task. When getting research done becomes effortless for us, we will naturally accelerate toward executing more and much more research and making the most of all the inventions that the world's advancement demands.

The more research work we complete, the greater the number, quality and value of inventions we make for the advancement of the world and the betterment of the future. Concentrate everything in you and about you in researching and making inventions upon inventions.

Ensure to succeed every day in living up to the firm determination and commitment device made. Even little daily writing of suitable ideas on your research, over time, compounds into something great in the science of research execution and invention making.

Note that merely writing a sentence, paragraph, page, chapter or section daily on making the target invention of your research is incomparably better than writing nothing about it. Daily executing a little unit of your experiment will compound into exceptional completion of your ultimate invention research goal over time.

Reading one page, chapter or section of a book each day cannot be compared to not reading a book at all. A minute of reading a book is substantially better than not opening a book at all. Daily journaling your research thoughts, no matter how short, is much better than refusing to write down your thoughts about the research.

Doing some of the experiments in a day is far better than failing to experiment that day. Practising research execution for even a few minutes daily cannot be compared to not doing it.

Analyze some data daily. Even successfully entering a column of data onto the statistical software interface or data editor is much better than doing nothing about the data analysis. Each little right action for completing the research matters in gaining more experience and accelerating closer to making the target invention. Work every day on completing the research until it is wonderfully done. It is the best daily contribution a person can make to global sustainability and advancement.

We live in a fast-changing world. Every change is directly or indirectly occasioned by research. Research produces the inventions that change people, society and the world for the better.

We must engage in research execution and be part of the people who originate, sustain and accelerate the fast-changing world for the best.

Research, build and own the better tomorrow we crave having. With passion, creativity, bravery and unreserved commitment, research and make the inventions upon which the future we utterly desire depends. We must do the research today and make the tomorrow of our dreams a reality.

Never break the chain of research, and you will end up with more inventions and discoveries that advance the world. Complete one research work in a grand style and embark on and accomplish another. Refrain from breaking the chain of researching, inventing, making possible impossibilities and turning unsolvable problems into solvable ones to improve humanity, progress society, and advance the world.

INVENTION-MAKING ACTION ALONE PRODUCES RESULTS

Invention-making action alone produces the desired results. Take action always to make the target invention. Every invention, creation, innovation, or discovery is a function of an ever-increasing action exerted by the inventor. The action that has to be taken is doing every bit of the work exceptionally well in executing the invention-making research fruitfully.

Plan and implement the making of such an astonishing invention that will overwhelm the world. The research plan, or research proposal, serves as an implementation plan. The research plan is an intention in writing that is made beforehand about when, where, why, and how to act and what to do to best accomplish an invention, discovery, creation, or innovation of remarkable worth.

The implementation plan makes it more likely for the exertion of all the needful actions that will bring about the target invention. Without an implementation plan for making an invention, a person is far less likely to get the invention made.

People who make a specific plan for what, how, why, when, and where to perform each research execution task are more likely to follow it through until the completion of the target invention. Preparing a concrete action plan, implementation intention, or research proposal is necessary for almost every invention-making journey. It is always better to embark on a very well-planned journey.

Then, follow the implementation plan to perform each predetermined task. One does not need to put off any planned action exertion for a later date. Perform each action at the planned time. This way, the actualization of the target invention becomes easier and surer.

Irrespective of how good the goal and the plan are, there can only be results with the taking of all the necessary actions. Implementing the masterplan for making an invention is indispensable for actualizing the invention.

Only the invention-making action, only the implementation, only completely dedicated exertion of ceaseless action delivers the outcome, the invention in view. The ultimate research goal can only be achieved by taking every needful action.

The degree of the right action exertion is proportional to the invention the research produces. Put little or no action in implementing the well-planned research; nothing, absolutely nothing gets invented, discovered, created, or innovated.

Superabundant exertion of all the needful actions results in making a fantabulous invention.

An innovation, invention, discovery, or creation can only be made to the proportion of action exerted in the execution of the research. Of the truth, there can never be research without the necessary action exertion. To research, action must consistently be taken on the investigation until the accomplishment of the target invention, creation, discovery, or innovation and its industrial production, mass commercialization, boundless dissemination, diffusion and extensive utilization.

"I will do this" and "I will do that" are mere proposed ideas. Proposed ideas alone never get anything done. People are unproductive, not inventive, and create nothing mainly because they operate only at the proposal level.

They never do the things proposed. They never do the right things they intended to do. They are only intending. They never take the requisite action to complete the inventions they wish to get done.

When they do not take all the mandatory actions for making the invention, they never accomplish any research. They never make an invention, innovation, discovery, or creation because of their inaction. Every needful action must be thoroughly taken before an invention can be made.

Inaction stagnates progress, hinders development, prevents invention, discovers nothing, and hinders innovation. Inaction retrogrades the world. The lack of invention-making action is speedily killing society and the world.

To invent, desperate, needful action must be taken ceaselessly until the target invention is exquisitely made.

Invention-making action exertion alone is what best makes the difference between inventors and the rest of humans. Many people who have not invented anything have wonderful ideas and plans of what to invent. But they never invent it due to their inaction and refusal to implement the plan.

It is with action on the research execution that inventions are made. Concentrate on taking all the actions to accomplish your research and the target invention.

A man with a thousand great ideas on what to invent but takes no action on any of them never invents a thing. Please, take ever-increasing action for an invention to be made. Take ceaseless action on each invention-making idea you have until the invention is wonderfully made.

Accomplish the research for making the desired invention. Do not merely be preparing to execute the research. Rather than being in motionless preparation, take every of the implementation actions to get the investigation completed.

Never remain merely at the planning stage. Instead, take every action to actualize the target invention, innovation, or discovery. Put the plan into action. Implement the various steps the research execution entails and get the invention made.

No one becomes an inventor merely by planning, proposing, wishing and hoping to do. It is by doing, exerting all the needful actions that a person invents.

It is with the making of several inventions that one becomes a prolific inventor. Far from planning and planning, proposing and proposing, take every practical action, and perform all the tasks for exceptional completion of each invention-making journey.

The research execution alone results in the target invention. Take the action now, and keep exerting it at an accelerated pace until a thunderstruck actualization of the target invention. Now is the auspicious time for executing the research.

Never hold yourself back from taking the implementation actions for the outstanding completion of your research in the pretence of planning, proposing and intending to do. Do the research now and get the invention-making journey completed.

Now is the best time for performing the various tasks to complete the research. Now is the auspicious moment of the research execution. Now is the most auspicious moment for taking action toward the dazzling completion of the invention your research aims at.

Motion without action never gets research done. Take all the necessary actions upon which making the invention depends. Be engaged in exerting action until the target invention's astounding accomplishment.

It is by executing research completely over and over, again and again, repeatedly, that a person becomes a prominent researcher and prolific inventor. A person must accomplish research repeatedly and make many inventions to be included in the top list of inventors.

Please make assertions like these and work until they are materialized:

- I have completed research very many times.
- I have made many inventions.
- I have become a veteran inventor.
- I am globally renowned for value creation.
- My inventions dominate the global marketplace.
- I am equated with my many momentous inventions, discoveries, innovations, and creations.
- My inventions have attracted me abundant wealth.

Habituate research execution. For researching to become a person's dominant habit, he must carry out research repeatedly. By doing research repeatedly, a person becomes automatic in performing key research execution skills.

The more a person repeats research execution, the more his

brain structure changes to become efficient and effective in research execution and invention-making. When an individual completes research, the connection between neurons in his brain for making inventions strengthens for considerably more research engagement.

With each repetition of research execution, the neural connections in his brain for making inventions improve, increase, tighten, and strengthen for the accomplishment of more and incredibly more research. Please, complete one invention-making journey astonishingly. Embark on and excellently complete another, and another, over and over, again and again, repeatedly. Then, the world shall celebrate you for your inventions, discoveries, innovations and creations.

The more research work an individual accomplishes, the greater the likelihood of his making more inventions, discoveries and innovations. Please, complete one research work and embark on and accomplish another. Keep executing research, one after another and become a renowned researcher and prolific inventor.

The more research a person completes, the greater research execution becomes his habit. When a person habitually executes research, he becomes far more proficient in research and valuable invention-making. The person makes more valuable inventions, discoveries and innovations.

Embark on research upon research successfully. Make more and much more valuable inventions. The frequency of inventions a person makes over time makes the difference between him and the rest of humans.

Do not relax in performing research and making inventions. Keep researching and inventing for the advancement of the world.

WORK UNRESERVEDLY HAPPILY TO INVENT

Unreservedly work happily on materializing your research goals, one at a time. Thoroughly practice and unreservedly work happily on completing each invention-making journey you embark upon.

It is with happiness that the various tasks or parts of research are successfully done. Happily, do every one of the things required to accomplish each aspect of the research you execute.

With displeasure and an unhappy heart, a person is not likely to make an invention of utmost value to humankind. To make an invention of great worth, the person must do every bit of the necessary research happily and pleasurably, irrespective of the challenge. The value and joy the invention to be made will add to humankind should motivate him to keep going on and on with exertion of extreme actions for the actualization of the target invention. The anticipated world's advancement should keep the investigator happy and satisfied while rigorously accomplishing each subgoal of the research until the marvellous actualization of the ultimate goal – having the target invention excellently made.

Never see or treat the performance of any task, the investment of anything, and the sacrifice of any kind as being too much for the making of the essential invention your research aims at. Happily investing time, talent and treasure in research is a prerequisite for making extremely useful inventions, discoveries and innovations.

To make an invention of inestimable usefulness with your research, you must happily invest all the requisite treasure, talent and time in it before nature will let you make the invention. Only when a person invests in and sacrifices incomparably more than what all else has put in for the making of a particular invention that nature permit him to get the invention made.

With enough committed investment of time, resources and talent, far more than other humans have done toward the invention of something with their research, no wall built of any material can separate you from successfully making that invention. Be practically determined to happily invest all you have in your research until the ultimate goal, making the target invention exceptionally, is accomplished, commercially mass-produced and disseminated. You automatically sing the victory song when the product is disseminated and makes remarkable waves everywhere.

SING THE VICTORY SONG FOR SUCCESSFULLY INVENTING

Research, invent and experience the feeling that you have conquered the world. Let your research make an exceptionally useful invention, creation, discovery or innovation. Then, you will ecstatically sing the victory song automatically:
Eureka!
I have found it.
Eureka! Eureka!!
I have conquered the insuperable.
Astonishment!
I have surmounted the problem.
Indeed! Indeed!!
I have overcome the world.
Thanks to God!
Nature has unveiled to me the ever-before concealed solution.
Greatness, glory, and wealth are mine!!
I have conquered the world.

On the accomplishment of the invention, I felt freedom, complete freedom. I felt ecstasy that I am in a new world. A totally different world – the world of inventors. I am now in the exclusively fantastic world of inventors. I am no longer like all other humans. I am now an inventor.

Next, orders were placed for the invented product weekly. Then, on a daily basis. And even in every hour from across the globe. Truly, I am a new person in a new world.

This is the world I must remain in, and I strive to get to the very top of it. Embarking on much more rigorous research is the key to making far more valuable inventions that satisfy the great needs of humankind.

In the new world, a new world of consciousness, the only thing that matters is making inventions and creating value. Investing every talent, treasure, and time in research execution and successfully making incredibly valuable inventions to best meet man's needs is all that matters. I must live up to the demands of the new world, the inventors' world.

There is nothing as good as making an invention of tremendous value. The eureka feeling and exclamation for having successfully invented, created, discovered, or innovated something of immense value for improving people's lives and advancement of the world will surely motivate the inventor exceedingly to execute more demanding research and make far more indispensable inventions.

When the person invests 50 per cent more than what it takes in yet another research, he will eventually make another invention. The invention shall turn possible another impossibility and solve an additional unsolvable problem to meet humankind's needs better. Yes. Such insuppressible motivation to research and invent more is what making an essential invention, creation, discovery, or innovation does to ensure an ever-accelerated advancement of the world.

CHAPTER 11

PRACTICE RESEARCH SKILLS TO AUTOPILOT LEVEL

The very best action for making inventions is intense, repeated, and prolonged practice. Intense repeated and prolonged practice of the various sections, aspects or parts of research execution is the best action for the astounding attainment of the ultimate research goal – the target invention.

1. Practice research execution skills to autopilot level.
2. Prevention of failure in research execution.
3. One practice plus another.
4. Practice during odd circumstances.
5. Become skillful in research execution.
6. Repeatedly practice invention-making skills.

PRACTICE RESEARCH EXECUTION SKILLS TO AUTOPILOT LEVEL

Long enough intense repeated practice of each aspect or section of research execution to autopilot performance level is the greatest key for making inventions. Autopilot expert performance of each of the steps in the execution of the invention-making is what best characterizes and sets prolific inventors apart from the rest of humankind.

Intensely practice the performance of each task in research to the autopilot level. Autopilot level of research task performance is the point of doing the tasks expertly automatically with little or no effort.

Repeatedly and intensely practice every aspect of research execution until it becomes part of your implicit memory. When the performance of a research section becomes part of your implicit memory, you expertly do the task unconsciously. You perform the tasks exceptionally well in an automatic form.

A person should practice an aspect of research execution to autopilot level. Autopilot level of performing an aspect of research is when, through intense and excessive practice of that aspect of research, it becomes part of the person's implicit memory. It is when the tasks for that section of research execution are done superbly by the person effortlessly. He gets the tasks for that research section performed wonderfully well without necessarily thinking about them.

Intensely practice repeatedly each part of the research execution to the point that you can do it effortlessly, just as eye blinking. At such a level of research execution, you become prolific in making inventions. You become like the celebrated prolific inventors in the world.

The more research tasks you practice to automation level, the more profound you become in research execution and invention making. The more automatic you become in expertly performing the various tasks in research, the more profound and prolific you get in making inventions, discoveries, innovations and creations that improve people and advance the world.

Repeatedly practice the various aspects of research execution to autopilot level, the level of implicit memory, at which you don't have to think about them before performing the tasks expertly. Then, you become renowned in research execution and invention making.

PRACTICE RESEARCH SKILLS TO AUTOPILOT LEVEL

With intense and excessive repetition of the various parts of research execution, recurring pieces of vital invention-making information get packed up and firmly stored in the implicit memory. This enables the person to attain autopilot expertise in research execution and in making inventions that turn impossibilities into possibilities and unsolvable problems into solvable ones.

Only persons who have enormously practised to the point of flawless and effortless automatic performance become renowned researchers and profound and prolific inventors. With absolute or complete commitment, any person who truly wants can intensely practice research execution repeatedly enough to become an adept researcher and a prolific inventor.

Excessively practice the different aspects, parts or sections of research execution intensely until they become permanently tattooed into your memory. Extremely practice and commit every part of research execution firmly into memory such that you can expertly do them automatically.

Expertise execution of research for making inventions involves a lot of complexity that only complete focus, dedication, and passionate repeated practice can lead to autopilot mastery. Relentlessly practice intensely to commit hundreds of research methods and designs, experimentation and objective observational approaches, statistical analysis techniques, and results dissemination procedures to memory.

Practice and expertly master the many aspects of research execution. Scaffolding the units or aspects of research practically before putting them together in real-life settings might help in easily reaching the autopilot mastery level.

According to Peter James Kpolovie (2023, 220) in **Research: Make Impossibility Possible**, "perfection in any aspect of research execution is a function of dedicated, deliberate, extended, excessive, extreme, relentless, resilient, and intense repeated practice."

The individual parts of research execution skills must be thoroughly practised to automate stepwise before synthesizing them into a single whole to make each invention. Exceptional practice and autopilot mastery of the various parts of and skills for research execution optimize invention-making performance in real time.

Direct your emotions in productive ways. With practice, change your emotions productively to ensure the best accomplishment of each research journey you embark on. Be emotionally aroused optimally for the best actualization of every research goal you embark on achieving.

PREVENTION OF FAILURE IN RESEARCH EXECUTION

When we practice a research execution skill enough, we gain full control over our response to and triumphantly overcome temporary setbacks and disappointment scenarios that a researcher encounters sometimes. Extendedly practising research skills keeps a person up and going all the time, no matter what, towards attaining the ultimate research goal.

Extended practice makes a researcher resilient and obsessive with the ultimate research goal actualization. It makes him never give up, irrespective of the challenges encountered during the invention-making journey. Extended practice makes him never doubt his abilities, skills and talents for outstanding accomplishment of the ultimate target invention. It makes him never get distracted from pursuing the target invention, even in the face of crisis.

A researcher is never defined by his occasional failures to hit the target but by his successes in making the invention. In case you have attempted research and failed once or twice to make

the target invention, do the following:
1. Know now that you are far better prepared to make that invention with additional determination, effort, sacrifice and practice.
2. Treat the failure as a first attempt at learning invention-making.
3. Now, you have learnt much better never to quit the invention-making journey until its exceptional accomplishment.
4. No completely committed effort toward one's research goal attainment dies abruptly.
5. Dedicated effort in pursuing the attainment of the invention aimed at by the research never dies without producing good results.
6. Such effort never dies until the realization of the goal.
7. Encountering failure is one of the milestones that a researcher overcomes on the path to successful invention-making.
8. That a person once failed does not in any way mean that he cannot succeed in making the invention.
9. It merely depicts that a person does not succeed every time in making the target invention at the scheduled time.
10. With much more determination, expertise and commitment, the person surely gets the invention made.
11. The person should practice much more, become more super-skilful in the research execution, develop his research execution skills to the fullest and exert much greater effort to accomplish the invention.
12. He eventually completes the research and gets the invention fantastically made.

Let me use some common sporting activities to illustrate the key points made. Whenever the best golfer misses the target, what does he do? He deliberately practices extendedly much more than ever before to super-perfect his skills. With better-perfected skills, he prevents or reduces the recurrence of such mistakes.

Similarly, when the greatest hitter in baseball misses a hit, he embarks on extended deliberate practice, much more than he has ever done to super-perfect his skills at the game. The same is true of professionals in the other various areas of sports.

Such extended deliberate practice for the super-perfection of one's skills is the greatest lesson in sports for researchers to learn from. People in other professions equally need to learn the great lesson of intense, extended, deliberate practice for super-perfection of one's skills from renowned sportsmen and women.

Learn the greatest lesson from sports professionals to super-perfect your research execution skills. When a person fails to hit the target invention in his research journey, what should he do? Like the professionals in sports, he should dedicatedly embark on much more intense deliberate practice of research execution skills than has ever been done before. Much more dedicated practice repeatedly for a long enough time is the surest approach to prevent failure in research execution.

Like sports professionals, a researcher who once fails to make the target invention should practice excessively to super-perfect his invention-making skills to guarantee his making the invention subsequently. Create and utilize every opportunity to practice much more, super-perfect your research execution skills and automate the skills for making valuable inventions, discoveries and innovations.

Kill procrastination. Procrastination is a deadly habit that most repels and prevents research execution. Procrastination is what prevents the making of inventions the most. Procrastination is a voice which is productive of every evil that prevents the making of inventions, discoveries, innovations, and creations. Procrastination is a self-destructive child, the mother of iniquities, and the father of all evils that stagnate and retrograde the world.

To make an invention, one must form the right habits, the habits of regular daily dedicated reading, writing, experimenting, analyzing data, and instant task completion. With the right habits, there is nothing of great value that a person cannot invent, discover, create, or innovate. Please habituate research execution and invention-making.

With intense practice, drill yourself until your research execution routines become automatic. Automate invention-making routines.

Form the right invention-making mindset and operate on the right attitude. Function with certitude that I am ready to do anything. Anything! Just anything to exceptionally accomplish the invention I crave. I must invest everything in the research and achieve the target invention.

Create new routines of what to do each night instead of sleeping or doing the things you currently do that have never allowed you to make inventions. Your daily new routines each night should be reading, writing, and experimenting. Each of these routines should be practised daily for a minimum of two hours.

When you practice these new routines for at least six hours every day consistently until they become your typical habits, you will be accelerating swiftly towards the attainment of every research goal you set. Routinely practice reading, writing, experimenting, and data analyzing every day. Practice these research execution routines to the level of automation – a point at which you do them exceptionally well effortlessly.

The making of valuable inventions should be the neurological reward that triggers habitual cravings in you. Making inventions should become what you primarily crave. Let making inventions become your cravings. The cravings should compel you to effortlessly demonstrate the invention-making routines of reading, writing, experimenting, analyzing data, and reporting. These new routines should replace the anti-invention-making behaviours that have hitherto characterized you.

ONE PRACTICE PLUS ANOTHER

Renowned researchers are inventors. Inventors are very rare people who make inventions, create and build things of great value for humankind, and the skyrocketing of the world's advancement. They devise and originate new and better processes, appliances, products, machines, ideas, articles, and services that rapidly improve and progress the world.

Please, become one of such persons. What is required to become a renowned researcher and prolific inventor is simple, easy and doable by anyone who truly cares to be.

Exceedingly doing one thing plus another consistently over time is demanded for a person to become a profound researcher and an outstanding inventor. The person must value two simple things, 'one' and 'plus'.

The power of 'one' and 'plus' cannot be overemphasized for an individual to become famous in researching and inventing. To become a prolific inventor, consistently over time, 'one' practice 'plus' another is necessary.

He must practice and attain mastery in each of the various aspects of research execution. That is, he has got to practice and attain super-expertise in 'one' research execution skill 'plus' another research execution skill and keep the process of mastering 'one' aspect of research 'plus' another aspect going consistently.

Super-expertise refers to expertise far beyond normal human capabilities in research and invention-making. Any person who dedicatedly practices research until he masters research execution far beyond normal human capabilities is bound to make inventions upon inventions, 'one' at a time, and become a prolific inventor over time.

He must successfully embark on research works, one at a time. He has got to make 'one' invention 'plus' another invention. In addition, he must keep accomplishing 'one' invention at a time 'plus' completing another invention at another time, going. He must practice and attain mastery in each of the several aspects of research execution.

'One' deliberate practice intensely 'plus' another 'one' daily consistently over a long enough period makes a person attain expertise in the specific research execution skills. When he keeps practising 'one' research execution skill 'plus' another daily committedly over a long enough time, he consistently improves in the timely completion of research until the attainment of super-expertise.

The power of 'one' practice 'plus' another practice, repeated over time, makes a person excellent in every skill and endeavour we can think of. Exceptional success in research and invention making, and indeed, in any desirable skill, depends on 'one' practice 'plus' 'one' practice consistently over a long enough time. Any research skill a person deliberately practices every single day for a long enough period, he maters it.

With engagement in consistently practising one research execution skill plus another, a person can achieve expertise in performing the skills over a long period. An individual must repeatedly practice a skill, tens of thousands of times, to become preeminent in it. To become distinguished in making inventions, a person must consistently practice research daily over a long time.

It is 'one' practice 'plus' another each day that, over time, collectively refines and improves an individual's invention-making skills to the peak. Sustained practice daily for months, years and decades makes a person prominent, highly honoured and widely acclaimed in his field. One practice of research upon another on every relevant occasion over a long enough time makes a person an eminent researcher and a prolific inventor.

There is never a skill for making inventions that a person practices consistently daily and thousands of times for years that he will not master. Whatever a person practices committedly for and long enough time, he attains a mastery level at it.

Please practice research execution and the making of inventions without relenting. You will eventually become an outstanding inventor when you do that consistently for a long enough time.

One becomes great in something by doing it over and over, repeatedly. The sustained, committed practice of reading, writing, experimenting, data analysis, patent drawings and application, and dissemination or reporting of the findings for many and long enough times keep a person improving research execution and invention making.

Please, always seek to attain super-expertise performance daily by getting slightly better in research execution. Continually refine and improve your research skills. Deliberately practice and learn to continuously improve how well you maximize your potential and achieve a super-expertise performance level in your invention-making journey.

Utilize the boundless power of 'one' and 'plus'. One alphabet plus an alphabet, persistently make a word. A word and a word

PRACTICE RESEARCH SKILLS TO AUTOPILOT LEVEL

consistently enough make a sentence. 'One' sentence plus another 'one' unswervingly makes a paragraph. A paragraph plus a paragraph makes one chapter. One chapter plus another chapter continually makes a book.

A book plus another book constantly, over time, make a library. Commercially publishing one book and another book, continually over time, get research findings preserved and disseminated for improving the lives of the audience, progressing society and advancing the world.

The power of 'one' and 'plus' another is inexhaustible in all fields of life endeavour. Consistently, over time, one penny plus another makes a dollar. Over time, making and saving one dollar plus another makes 100 dollars.

Over time, a hundred dollars plus another hundred dollars unswervingly make a thousand dollars. One thousand dollars plus another constantly makes a million dollars. Keeping a million dollars and a million dollars steadily will make a billion dollars over a long enough time.

One dollar is incomplete without a cent. A million dollars is not complete without 'one' dollar. Every cent and every dollar is of great importance in wealth creation. So, it is in the making of inventions. Every invention-making skill exceptionally mastered is essential in getting an outstanding invention made.

Deliberately, intensely and committedly practising one research execution skill plus another every single day for many long enough days, weeks, months, years, and decades makes a person proficient in research completion and invention-making. Proficiently executing research, one after another over and over, again and again repeatedly, culminates in inventions, discoveries, innovations and creations of value.

Consistently making 'one' vital invention 'plus' another great one over time makes a person a renowned and prolific inventor.

Making one invention plus another unswervingly over time advances the world beyond imagination.

Every good thing begins with one. When one good thing is steadily added to another, it compounds into a great thing. Inventing 'one' great thing of value 'plus' another routinely for a long enough time, people's lives improve, society progresses, and the world remarkably advances.

Never stop making improvements in research execution. Keep increasing the number and value of your inventions, discoveries, creations, and innovations. Continuously advance the world with an additional invention, creation or discovery.

Never stop learning. Keep improving your research execution expertise. Continue to execute research one after another successfully. Continue to make 'one' invention 'plus' another. Constantly create remarkable fortune for people, society and the world with your inventions, one at a time.

PRACTICE DURING ODD CIRCUMSTANCES

Always demonstrate bottomless passion in self-improvement for making much more valuable inventions. To make your next miracle invention, work daily when it is convenient, exciting, and motivating and when it is not. Work, particularly when the circumstances are very odd and challenging.

Every person can work hard when he is on a task that is immediately exciting, motivating and convenient. What will easily set a person apart from the rest of humans is when he continues to accelerate exertion of an ever-increasing action and effort in his research particularly during moments that are not immediately gratifying, not convenient, not exciting and not motivating.

Working with increasing enthusiasm when it is odd and most

PRACTICE RESEARCH SKILLS TO AUTOPILOT LEVEL

challenging sets a person apart from the rest people in his invention-making journey. Never a day let boredom and task difficulty prevent you from practising and improving your research and invention-making skills.

No matter the tens of thousands of times you have practised in the past, continue to increase your practice of invention-making skills. The more you practice at odd times, the more you overcome boredom in practising. Plus, the more you practice during odd times, the closer you get to attaining super-expertise in research execution. The greater the practice in odd circumstances, the closer you reach maximum potential in any professional field.

Scottie Scheffler, a leading golfer; the best tennis players, such as Carlos Alcaraz and Novak Djokovic; the top basketball players – Nikola Jokic and Stephen Curry; the greatest footballers – Lionel Messi, Kylian Mbappe and Kevin De Bruyne, never stopped practising their skills even after they have practised millions of times.

Similarly, the top prolific inventors and renowned researchers in the world (United States Patent and Trademark Office, 2024), such as Shunpei Yamazaki, Kia Silverbrook, Thomas Edison, Nikola Tesla, Robert Fulton, Lowell Wood, Gurtej Sandhu, Melvin De Groote, Donald Weder, Leonard Forbes, Kangguo Cheng, Philo Farnsworth, and Jerome Lemelson, all kept practising and perfecting their research and invention making skills daily, irrespective of how numerous they have individually practised in the past.

Even after practising research execution tens of thousands of times, they continued to practice the skills more and more, particularly when it was immediately inconvenient, not gratifying, unexciting and not motivating to practice. Please practice research execution, particularly during very odd moments and circumstances. Only the practice done at very odd times and circumstances that count in getting a person to the top of his profession.

The more a person practices research execution skills, the better he becomes at making inventions and the more difficult it becomes for others to compete with him. The greater he practices research execution, the closer he gets to the top list of inventors.

Daily practice the skills for research and making inventions much more than the great inventors have done, far beyond what people in their wildest imaginations would expect of you. Practice principally during odd circumstances and attain the peak of your research and invention-making potential.

Exert the unique efforts required to reach and remain at the maximum of your invention-making and creative potential. Practice, master, improve, and work much harder than the best inventors you admire. Execute more research than they have done. Make much more valuable inventions than the best inventors you have read about.

Our genes and talents do not eliminate the great need for repeated intense practice and hard work under very odd circumstances to succeed in research and making inventions. No matter how gifted, talented, or intelligent a person may be, it demands the person's ceaseless repeated practice at odd moments to attain super-expertise in any given field. It takes the person's intense repeated practice of research skills and hard work at research completion, particularly during very challenging circumstances, to succeed in making inventions, innovations and discoveries.

Talents and genes without deliberate, committed, prolonged, and excessive practice and hard work, particularly under the most challenging circumstances, a person cannot research and make the inventions that will move him into the list of top inventors. A person must work very hard on and countlessly practice experimentation, reading, writing, data analysis, patent drawing and filing, and mass production and dissemination of the findings before he gets into the top list of inventors.

Therefore, a person must excessively practice and work exceedingly hard and smart under the most challenging circumstances in researching and making inventions of immense value. Emphatically, remarkable inventions are only made by a person who researches when it is inconvenient, not motivating, unexciting, and extremely challenging. Work at your alarm zone to make inventions of utmost value.

BECOME SKILLFUL IN RESEARCH EXECUTION

Develop research execution as a skill by practice until making inventions becomes your special talent. Invention-making journey, regardless of how far it may be, can be fueled to the attainment of the cherished goal. Unshakable self-confidence, irrepressible persistence, and immeasurable love for making the target invention are the unfailing fuel.

Boundlessly love and cherish the making of inventions far and above all things. With enthusiasm, make extraordinary sacrifices for and extreme investment of all the needful effort, talent, time, resources, and action in making an invention until the realization of the ultimate goal.

When the making of an invention is the one and only thing you love the most, you can effortlessly exert all the needful extreme actions to achieve your target invention. At each time, be enthusiastically engaged in the making of the invention you love the most until it is fantastically accomplished.

Each invention accomplished increases the inventor's self-esteem and better repositions him for more creative and inventive exploits. The improved self-esteem and motivation allow him to embark on subsequent invention-making journeys with more confidence, expertise, and optimism. These increase his chances of successful timely completion of every of the research journeys.

Every invention, discovery, innovation, or creation made increases the researcher's experience to do something much more valuable, challenging, creative, nurturing, and demanding to advance humanity. It moves him speedily closer to his maximum potential attainment.

Each invention a person makes reinforces his self-confidence and inspires others' continued confidence in him. When we make inventions, the world responds kindly to us, acknowledges our breakthroughs, and better respects us. These increase our worthiness.

Every invention a person makes with his research automatically increases his research execution skills beyond imagination. As a person records more inventions, research execution becomes his habit. Successful execution of research over and over again gets a person to habituate research execution skills.

REPEATEDLY PRACTICE INVENTION-MAKING SKILLS

Repeated practice is an indispensable rule that must be adhered to before a person becomes globally renowned in anything. A person can only be famous worldwide for what he has over-practised and done exceptionally well repeatedly often countless times.

An individual can only become a champion globally in what he has done or practised tens of thousands of times. Never a skill that a person is famously known worldwide for, that he has not practised, exhibited, and displayed countless times.

To become distinguished globally in any profession, a person must practice the skills the profession demands numerous times. Preeminence in any field demands countless practices of that endeavour.

Consequently, prominence in research demands the successful execution of research numerous times. Every renowned researcher must have successfully embarked on research very many times.

A prolific inventor is someone who has researched many times and made several inventions. Dedicate your life to research execution and invention-making. Make many valuable inventions, discoveries, and innovations to become renowned.

A person does not plant a tree and becomes famously known as a farmer. By building one house, an individual does not become a renowned estate developer.

You do not make one book publication and become renowned worldwide as a prolific author. Similarly, one research work does not make a person a prolific inventor.

A person must own large estates to be globally known as an estate developer. To become a renowned prolific author worldwide, one must write and publish many books. To become a prominent or outstanding inventor worldwide, a person must execute many research works and make several essential inventions.

Do everything it takes to become a renowned researcher and profound inventor worldwide. Get completely committed to researching and becoming a renowned inventor globally.

CHAPTER 12

BRAIN BOOSTING FOR MAKING INVENTIONS

The brain can and should be deliberately booted for research and the making of substantially more inventions. The human brain is highly flexible. Hence, every person who truly wants can committedly rewire their brain to function exceptionally well in research execution to make extremely useful inventions, discoveries, creations and innovations that greatly advance the world.

1. Boost your brain to make inventions.
2. Memory improvement for making inventions.

BOOST YOUR BRAIN TO MAKE INVENTIONS

Whatever could positively contribute measurably to increasing and keeping a person's brain healthy and functioning optimally is of some help to improving the person's capacity for successful research execution and the making of indispensably useful inventions for the improvement of humanity and the world's advancement. Outstanding research accomplishment demands the investigator to do everything necessary to boost his brain.

Neuroplasticity of the brain allows anyone to attain expertise in research execution and invention-making easily. Neuroplasticity is a treasured feature of the human brain that enables a practically desirous person to develop and become a profound researcher and renowned inventor. Take every necessary step daily to transform your brain and become renowned for research and invention-making.

Be not deterred. Keep learning, practicing, and experimenting to improve your brain daily for outstanding research execution. Do everything necessary for becoming superb in research execution, invention-making, and turning impossibilities into possibilities and unsolvable problems into solvable ones, as emphatically reiterated in this book, **Research, Invent, and Create Wealth**, and in an earlier book titled **Research: Make Impossibility Possible**, by the current author, Peter James Kpolovie (2023).

Study intensely to become good at getting information and permanently storing it in your brain, memory or head for use when and how needed for the successful making of inventions. With determination, hard work, commitment, dedication, and relentless practice, a person can consciously direct his brain and force it to stretch, bend, develop and grow to become super excellent in research execution and invention-making.

The process can take years just as several years of committed training are indispensable to attain expertise in each profession. Imagine the number of years of dedicated learning, practice and training to become an excellent pilot, surgeon, AI scientist, Software developer, neurologist, cardiologist, data scientist, biochemist, optometrist, industrial engineer, or information security analyst.

Never a current technology that is as nimble as the highly adaptable and marvellously flexible human brain. Improve your brain to the fullest for making discoveries, innovations, creations, and inventions that your research targets.

Neuroplasticity of the human brain characteristically enables it to reshape and adapt its way of operating. Hence, anyone who dedicatedly chooses to stuff his brain to become like that of a remarkable researcher and prolific inventor can achieve it.

The brain is not constant. With determination and dedication, the brain can be trained and improved upon greatly to engage in research for the advancement of the world.

The status of the world today and in the future is subject to or dependent upon the magnitude and volume of research that we train our brains to execute impeccably. When you are committedly determined to successfully embark on and accomplish research work by inventing something of great worth, the world changes for the better with one indelible leap.

The choice is completely yours to exceptionally develop your brain and successfully embark on research for making essential inventions, creations, discoveries, and innovations upon which the advancement of the world depends.

On the contrary, if you choose not to train your brain for research execution, you can't invent anything. Consequently, you only impede people's development and retrograde the world. Impediment of people's advancement and retrogression of the world are never what you or any person should do.

When a person chooses not to research and refuses to fund others' research, he can never invent a thing. Then, he fails to accomplish the golden mission for which God, the Almighty, created the person in his own image after his likeness.

The consequence of a person's refusal to embark on research directly or indirectly is retrogression of himself, people and the world. He causes the impoverishment of himself and humankind. Never retrograde or impoverish yourself, humankind and the world.

Suppose you know someone who had chosen not to research

and not to sponsor others' research before you read this book. In that case, I solemnly implore you to persuade the person to actively engage in more positive thoughts and actions that will cause brain alterations for changing his choice. When he changes his choice, he can execute research successfully or fund others to make essential inventions to advance the world and enrich humanity.

The best way to persuade such persons to change their minds is by acquiring and gifting them copies of this book (**Research, Invent, and Create Wealth**) and the earlier book by the current author titled **Research: Make Impossibility Possible** (Peter James Kpolovie, 2023). When they read these books, they will most likely have a rethink and make the right choice of embarking on research execution and or funding other people's research for the making of inventions and creation of value.

Each copy of these books a person acquires is a remarkable way of funding others to research and make inventions. Proceeds from the books are reinvested in invention-making and further motivation of people to research, make possible impossibilities and turn solvable unsolvable problems.

We have a responsibility to change this world for a much better. Only by researching or funding others to do so can we invent the things a much better world depends on. Please, we should never grow weary of doing the needful for people's improvement and advancement of the world.

Through affirmative positive and deliberate actions, a person can cause the neuroplasticity of his brain to favour engagement in research execution or funding others' research for the making of valuable inventions and discoveries. An individual can easily take control of his brain and overhaul his life to research or fund research to make immensely useful inventions. This way, he easily fulfils the inventive and creative life mission for which he was made in the image and likeness of God, the creator.

The fact that you acquired this book and have read it up to this point demonstrates your committed engagement in research execution for making the inventions that having a much better tomorrow depends upon. Additionally, it is already key evidence that you are funding research and encouraging others to make inventions and discoveries of immense value. Congratulations. Please, keep it up.

A person should deliberately challenge his brain with all sorts of mental gymnastics to super-build the brain for invention-making, just as daily walking firms up the body for better physical looks. Regular dedicated mental training and practice can alter the brain circuit organization to enhance his overall brain power for research execution to make inventions and discoveries of great value.

Throughout a lifetime, the human brain remains pliable. Therefore, every person, irrespective of age, who deliberately empowers his brain to make inventions with his research can successfully achieve his goal. Every person's brain can better be restructured to become hyperintelligent for making inventions if he commits to the meticulous practice of research execution skills.

Deliberately cause the neuroplasticity of your brain for thorough research execution. It will result in the making of highly valuable inventions, discoveries, creations, and innovations to improve people's lives, better society and the world's advancement.

When a person extendedly practices and does everything necessary to transform his brain to become superb in research execution, invention-making and turning solvable unsolvable problems, he eventually becomes a preeminent researcher and prolific inventor. When a person exceptionally practices research execution skills many enough times and over a long enough period, a corresponding change in his neuroplasticity occurs for making inventions, discoveries and innovations to advance society and the world.

The intense rehearsals of research execution skills build the person's brain muscles in the making of much more useful inventions that add value and create the desired future we crave. Never ever cease practising research execution skills for the continuous reshaping of your brain to most effectively and efficiently engage in the making of essential inventions.

Whenever the brain is committedly challenged for enough and long enough with mental gymnastics, the brain cells, muscles, neurons and related nerves develop more formidable and tough. Thus making the brain much more capable of successfully embarking on highly valuable invention-making journeys.

The brain-building system for better research execution is analogous to regular daily physical training in strengthening the body for better looks. The better developed brain is much more capable and readily willing for research execution, invention-making, turning possible impossibilities, and making unsolvable problems solvable.

Every time a person dedicatedly thinks research execution thoughts, feels invention-making emotions, and performs actions for research accomplishment, there is always the release of neurotransmitters that result in some sort of plasticity in his brain. Such plasticity better shapes the person's brain to become ultrasmart for attaining the ultimate research goal – the target invention accomplishment.

As you read this book, your brain is much better restructured and strengthened for research execution and invention-making to advance the world. Almost everything that could be done to markedly improve a person's brain for making inventions, discoveries, innovations, and creations plays some role in enhancing his research execution capacity. Continue to improve your skills for research execution and keep accelerating closer to making many more inventions.

MEMORY IMPROVEMENT FOR MAKING INVENTIONS

Memory plays a fundamental role in the mastery of the various research execution skills to the autopilot expertise level. Never a research skill that can be performed automatically without first practising it intensely and committing to memory.

Before a research execution skill can be mastered to the autopilot level, it must have been previously practised repeatedly and firmly committed to memory. When an invention-making skill is so well committed to memory, it becomes a habit that is effortlessly demonstrated whenever needed.

Memory plays a great role in the performance of every research execution action. Memory is a special faculty of the mind by which information or experiences are processed, rehearsed, encoded, stored, and firmly retained over time. Next, the information is retrieved whenever it is needed to build new knowledge to strategically improve both present and future performance. The retained information is retrieved as and when needed for the purpose of accomplishing research tasks.

Memory is the store of an individual's total rememberable experiences that uniquely distinguishes him from others in proffering workable solutions to problems of humankind.

The creation, invention, discovery or innovation of things, ideas, solutions, products, and services for the improvement of the future is a function of the magnitude to which a person restructures his stored information or past experiences and applies them better to meet the needs of humankind through his research. Thus, without memory, research cannot be done, and invention or discovery cannot be made.

The identification of a pressing persistent problem to research on and create, invent or discover a useful solution to is a function of the extent to which the person restructures his stored information or past experiences and practically applies them in new settings or circumstances.

Every one of the numerous tasks that the researcher must perform in executing the research for the successful invention of an exceedingly valuable solution to the problem all depends on the magnitude to which he restructures his memory, his stored information and past experiences. Therefore, having an excellent memory is indispensable in research execution for the making of extremely useful inventions or discoveries.

Memory is analogous to a well-recorded videotape containing all of a person's life experiences in such a flexible manner that it almost automatically restructures to directly influence all his research-related thoughts and actions at the present for the invention or creation of a much better tomorrow. A person's memory does not only store past information and experiences. Still, it lets him imagine a better future and what as well as how to invent the things that the materialization of the future depends on.

The imagination or visualization of a better tomorrow and the execution of the experiments that will not only stimulate and predict the improved future but actually make the advanced future a reality largely depends on the extent to which the investigator possesses and restructures his memory for the purpose. Memory plays a crucial role in predicting the future and executing experiments or research to actualize the desired prediction.

Exposing the memory to as many as possible new experiences is a great way of improving memory. Exposing the brain and memory to new experiences motivates and rewards enhanced learning via the role played by dopamine.

Experiment with something new and different from time to time to solve currently unsolvable problems and turn possible things that are currently perceived to be impossible. The new experiments you do provide your memory with new experiences and information that get integrated with already existing experiences and stored for further restructuring and use as needs arise.

Practice, learn and stuff research execution skills into your memory. Such learning makes a person more productive. It surely reduces the time he will spend hunting for the needed information to make inventions.

Learning, practising and stuffing many research execution skills into memory to the autopilot level enables the person to always focus single-mindedly on making the target invention in each research journey. It makes him consistently consider the consequences of his potential actions and constantly adjust the strategy accordingly to guarantee a most awesome accomplishment of the ultimate research goal.

Thoroughly practice each research execution skill until it gets permanently embedded in your memory and brain like an immovable rock. Attach a new research skill you encounter or are rehearsing to already unforgettably stored information in your brain. When a new research skill is embedded to firmly existing information or knowledge, it becomes more likely to retain the new skill in the memory for automatic use when the need arises.

Deeply encode new research execution skills by repeatedly practising until they become your typical characteristic that can very quickly and automatically be put into use when needed. Intensely practice the new research skill until it becomes your automatic habit.

Resoundingly practice each research execution skill until it is permanently ingrained in you so well that you can practically perform it perfectly even when asleep. Subject yourself to tens of thousands of hours of research execution training. When a person trains himself that much in anything, he surely attains expertise at it. Each time you read this book; you are exceptionally undergoing research execution training.

Practice research execution tens of thousands of hours and become prolific in the making of inventions, discoveries, creations, and innovations. Your creations, inventions, and discoveries shall automatically become an endless stream of uncommon wealth to you in addition to improving people and advancing the world. Indeed, one cannot crave anything better than making highly treasured inventions.

Thousands of hours of dedicated practice or training on research execution skills get you closer to making essential inventions for improving people's lives and advancing the world. The more intense the practice of a research execution skill, the more automatic performing the tasks becomes and the greater it contributes to the success of your invention-making journey.

Memorization or rehearsal of a research execution skill does not have to be in a real setting always to help you gain practice. Despite the fact that a person has not actually made an indispensably valuable invention, simulated training on the skills for making such an invention until the skills become part of his second nature gets him closer to the making of substantially useful inventions.

Mentally simulated research experience can occasionally serve the purpose as much as a real one. Deeply meditate, visualizing the invention-making scenario you want to rehearse. Focus all your senses on doing the research and making the invention.

Form a memory of making a necessary invention under highly emotional circumstances. In such highly emotional circumstances, the memory of invention-making skills is often recalled with much distinctive clarity and applied to making other inventions.

Highly emotional invention-making skills and experiences tend to be remembered more richly and applied subsequently than those for emotionally neutral experiences. Often, heightened emotional research execution skills cause memories to imprint deeply. Profound emotion could positively affect the memory of the associated research skills whenever the need arises subsequently.

Meditation, yoga and hard work

Meditation could help the brain in the generation of invention-making vision. Regular meditation helps alter the brain of an invention-oriented person for better research execution. With regular meditation, the structure of the cortex could be reshaped. And the thickness of the regions concerned with attention and information processing could be increased (Lazar, Bush, Gollub, Fricchione, Khalsa, & Benson, 2000).

For exceptionally successful research execution and invention-making, regular practice of meditation is a worthwhile pursuit. Every prominent researcher or prolific inventor devotes some minutes daily to meditation practice. Learn to do the same if you are completely determined to embark on research and accomplish essential inventions, discoveries, innovations, or creations.

Daily meditation practice, yoga and cognitive behaviour therapy rehearsals for some months could bring about measurable changes in the brain for better research execution. As demonstrated by Lazar, Kerr, Wasserman, Gray, Greve, Treadway, McGarvey, Quinn, Dusek, Benson, Rauch, Moore and Fischl (2005), regular meditation and yoga practice could increase lasting cortical thickness.

Structured meditation practice for even just a few months of mindful minutes daily could have a positive carryover into the rest of the person's life and in relatively permanent changes to his brain for improved research execution. With high determination and self-awareness, reshape your brain through meditation, yoga and hard work for successful research execution. Use meditation and yoga techniques to promote mindfulness and well-being that change the brain for more passionate engagement in invention-making research.

Hard work, particularly in regularly reading widely, committed intense experimentation, and dedicated writing, could improve the brain for embarking on successful research completion. Read all you can get your hands on about research execution, making inventions, innovations, and discoveries. Read all you can get about researchers and inventors. Join research associations so you can speak one-on-one to profound researchers about how to experiment and successfully invent the most valuable products, services, processes and ideas.

Think positive research thoughts. Everyone who cares can make a valuable invention, discovery, or innovation with research. Unfortunately, however, some people think negative thoughts, such as:

"I can't invent."

"I can't make a discovery."

"I cannot make innovations with my research."

Such persons should banish their negative thoughts and replace them with positive research thoughts. Banish all negative thought patterns. Develop positive research thoughts. Examples of positive research thoughts that a person should always be having include:

"I can invent."

"I can innovate."

"I can create."
"I have created great value for humankind."
"I have invented a crucial product."
"I have created an essential device."
"I have made an indispensable innovation."
"I have originated life-changing and world-improvement ideas."
"I am rendering unique service to humankind."
"I can make valuable discoveries with my research."
"I am practically exerting all the needful invention-making actions."
"With God in me, I can invent or create every good thing via research."

Positive research thoughts can reshape the brain and reorient life for outstanding research execution and invention-making. Developing positive research thoughts and restructuring the brain for invention-making is simple.

Devotedly engage in meditation practice and cognitive behaviour therapy rehearsals that could make the person better in research execution. Plus, unreserved exertion of experimentation actions for research completion.

The power of neuroplasticity that the brain has, allows the human brain to easily reshape itself all through life for exceptional invention-making directly through research execution or indirectly via funding others' critical research. The brain's neuroplasticity can be enhanced through intense and regular practice, yoga, meditation, hard work, and cognitive behaviour therapy. Get committedly engaged daily in:

1. Extraordinary practice of research skills repeatedly.
2. Extended regular yoga on invention making.
3. Profound meditation on the making of inventions.

4. Cognitive behaviour therapy for positive invention-making habits.
5. Remarkable experimentations.
6. Conscientiously break the ultimate research goal attainment tasks down step by step.
7. Next, sequentially accomplish each task unit from the most decisive to the least until the amazing completion of the target invention.

This way, a person can marvellously achieve the ultimate research goal with ease in every invention-making journey he embarks upon. Diligent performance of the above listed activities could help reshape and sharpen a person's brain for researching and making essential inventions. Excellently accomplish one research execution after another and attain your life's utmost mission and maximum potential.

At each period in a researcher's life, his brain exhibits an amazing restructuring to reshape itself. The neuroplasticity power of his brain makes the brain adapt best to research execution for the making of indispensably useful inventions, creations and discoveries.

Proven deliberate techniques for facilitating and enhancing the neuroplasticity power of the brain include intense, regular, long enough repeated practice of research execution skills. Others are as stated earlier, committed and dedicated regular meditation, yoga and cognitive behaviour therapy.

Additional methods for enhancing the brain exceptionally well for research execution include:
1. Eating brain-boosting foods
2. Having good sleep
3. Engaging in neurobic exercises

4. Continuous daily improvement of invention task performance
5. Doing a myriad of experiments.

Eating Brain-Boosting Foods

Eat brain-boosting foods to enrich the brain for excellent performance and research execution. Proven brain-boosting foods are those very rich in essential fatty acids. Foods loaded with essential fatty acids that the brain should be boosted with are primarily those that give omega-3 fatty acids (Kpolovie, 2012; Kpolovie, 2011).

Foods rich in essential fatty acids are critically essential for the building and sustenance of the super brain needed for successful research execution. Essential fatty acids are the main determinants of the brain's neuronal membrane fluidity. The synthesis and functions of many of the brain's neurotransmitters depend on the adequacy of essential fatty acids in the brain.

Essential fatty acids that the brain greatly needs are chiefly omega-3 essential fatty acids, without which the brain cannot function optimally in research execution and invention making. Eating cold-water fish such as salmon, mackerel, anchovies, seabass, oysters, sturgeon, sardines, herring, lake trout, shrimp, and albacore tuna gives the brain the necessary essential fatty acids.

Foods rich in omega-6 essential fatty acids and those rich in antioxidants are also helpful to the brain cells which could facilitate mastery of research execution skills. Eating foods rich in omega-6 and antioxidants helps protect the brain cells from damage (Kpolovie, 2012; 2011).

Sources of omega-6 essential fatty acids and antioxidants include pumpkin seeds, walnuts, soybean oil, seaweed and algae, chia seeds, flaxseeds, hemp seeds, wheat germ, kidney beans, and edamame. Others are pine nuts, sunflower seeds, corn oil, sunflower oil, walnut oil, grapeseed oil, blueberries, spinach, artichokes, and orange vegetables like carrots and sweet potatoes. Dark green vegetables are also very useful as they are highly rich in antioxidants.

The brain needs to be enriched for successful research execution. Omega-3 essential fatty acids, omega-6 essential fatty acids and antioxidants play crucial roles in the brain's enrichment, development, and optimum functioning (Kpolovie, 2012; Sade Meeks, Kat Gal & Charlotte Lillis, 2023).

Good Sleep

Having good sleep, both non-REM and REM sleep, could help enhance the brain's performance in some research-related tasks. Some hours of sleep each day tend to relax the brain and charge the brain to function a little better in the pursuit of an invention-making journey. Sleep, be it non-Rapid Eye Movement (non-REM) or Rapid Eye Movement (REM), could improve memory cell regeneration and memory that impacts research task performance.

During sleep, glucose metabolism in the brain increases, enhancing short-term and long-term memory for overall learning of research execution skills. Sleep helps in cell regeneration, repair of tissues and bones, and restoration of energy that could be exerted in task completion and research execution.

Some research tasks are better actively done after waking up from sleep. Enough sleep is needed to relax, restore, refresh, and revitalize the brain for the sharp completion of some research tasks. Occasionally, the hypotheses postulated just after waking up from sleep tend to be more testable for resolving some difficult problems (Kpolovie, 2011).

Sometimes, invention ideas are more easily captured in drawings or picture format just after waking up from sleep. More often than not, patent drawings are made within the first hour after waking up from sleep than other periods. To pictorially illustrate, elucidate and describe a new invention or a prototype of a new invention, something completely novel and unprecedented, the first few hours after waking up from sleep is the best moment.

The making of an invention requires patenting. Patent drawings that best capture the invention epiphany in view have to be submitted to the patent office for inspection, evaluation and approval. Every developed and a few Third-World countries have a patenting office for the purpose.

In the United States of America, the United States Patent and Trademark Office (USPTO) is in charge of evaluating and approving patent applications. The World Intellectual Property Organization (WIPO) is used for filling patent applications internationally. The Japan Patent Office (JPO), Canadian Intellectual Property Office (CIPO), European Patent Office (EPO), United Kingdom Intellectual Property Office (UK-IPO), and Russian Federal Service for Intellectual Property, and the India Office of the Controller General of Patents, Designs & Trade Marks (CGPDTM) each requires seeing unquestionable patent drawings before approving a patent application.

The approval can only be given if the patent drawings explicitly contain accurate details of:
1. The nature of the invention.
2. Functions of the invention.
3. How the invention operates in the real world.

The researcher often makes patent drawings from the moment he gets the invention-making epiphany. Invention-making epiphanies tend to occur most during the first few hours after waking up from sleep. The drawings for making the invention guide the researcher in designing the invention and working on it persistently until the actualization of the invention.

The first few hours after sleep are when a great number of invention drawings are made. Most often, even when such drawings about an invention-making idea are made outside one or two hours after sleep, the finetuning of the drawings is done within the first hours after sleep (Kpolovie, 2011).

Neurobic Exercises

Neurobics are mental activities and cognitive exercises that, when regularly practised as part of a person's daily routines, could help boost memory, stimulate the brain, prevent memory loss, increase memory recall, protect against cognitive decline, and stimulate the brain's muscles. Neurobic exercises regularly practised daily could collectively help in improving the performance of some research execution tasks.

Neurobics are mental exercises that appreciably sharpen concentration. With neurobic exercises, the brain's cognitive maps are strengthened, and memory of research execution skills is improved.

With sustained regular neurobic exercises, the brain muscles are stimulated and developed, and neurons and neuronal networks are optimized for research execution. Neurobics improve the nervous system, attention capacity, and perception acuity and noticeably boost the power of the brain. Neurobics could boost intellectual ability, intelligence, learning ability, and research performance capacity.

Neurobics are mental tasks and activities designed for brain stimulation and memory loss prevention. Regular neurobic exercises over time are capable of improving the mindset for invention-making.

Simple examples of neurobic exercises include the following.

1. Reading aloud ultra-fast and ultra-slow.
2. With eyes closed, walking to the door, getting the bunch of keys, picking the right one, and inserting it into the keyhole to lock and unlock the door from inside and outside.
3. Drawing a clock while looking at its reflection in a mirror.
4. Inputting a novel script on the computer from the mirror reflection of the script.
5. Using the non-dominant hand in writing, eating, playing a musical instrument, and brushing the mouth.
6. Learning and speaking new languages that requires using different parts of the tongue, lips, mouth, and vocal cords.
7. Counting one's breath while eating in the dark or with eyes closed.
8. Chanting aloud a series of research execution skills, statistical equations and syntax while taking a shower with eyes closed.
9. Taking different routes to and from work.
10. Pinching the ears with crossed arms and squatting for a few minutes.

Continuous daily improvement in invention task performance

Continuous daily improvement and progress with nonstop action toward making the target invention help to achieve the invention-making goal over time. Keep improving and advancing daily in performing the tasks required for the making of the target invention. With the nonstop daily refinement of your invention tasks performance, the target invention gets made over time.

Daily completion of research task units is a sure strategy for getting loads of rich and meaningful research experiences. Create or invent things of great value rather than merely consuming or using others' inventions. Earn money with your inventions rather than just spending on others' invented things.

Always prioritize research accomplishment over immediate comfort. Committedly work without relaxation to accomplish each research or invention-making journey exceptionally well.

Every extremely useful invention made with research, like each exceptionally great achievement in life, is the product of repeated small beginnings. There is never an overnight success in research and invention making.

Research execution mastery demands patiently working hard and being smart at it. The most indelibly useful inventions are made after long delays when persistent hard work is continually done.

The delays could be in months, years or even decades. If all through the delay period the researcher committedly worked dedicatedly nonstop at it, then, all of a sudden, the essential invention gets made.

Invention-making with research is much like the wonder plant, Chinese Bamboo, which takes five years of dedicated nursing from cultivation to grow out of the ground. When the bamboo is planted, it demands diurnal fertilizing and watering of the ground for as long as five years before it breaks through the ground. But after breaking through the ground (that takes five long dedicated years of watering and fertilizing), the wonder plant grows as tall as 90 feet within six weeks. To make inventions, one must work like the Chinese bamboo farmer.

To successfully research and make inventions, a person must practically resolve to continuously improve little by little, each day, in the performance of invention-completion tasks. Get better a little every day in research task implementation. Then, after a long period, the research gets done exceptionally well, and the target invention is outstandingly made.

Daily practice and improve your research accomplishment skills. Plus, execute the research skillfully with all amounts of dedication. Work continuously on the research every single day until the accomplishment of the invention-making journey in a fantastic manner. Continuous daily improvement and progress with nonstop action toward making the target invention, over time, achieve the invention-making goal.

Keep improving the application of your research skills daily. Daily exertion of an ever-increasing action toward the accomplishment of the research execution journey leads to the attainment of the ultimate research goal – making the target invention essential for improving lives and advancing the world.

Learn, practice and automate more and more research performance skills little-by-little every day. Apply the skills in embarking on life-changing research.

Break the research execution tasks down into small units. Prioritize the task units from the most important to the least. Plus, exceptionally perform each of the task units to completion.

Accomplish each of the task units, one after another, as prioritized. Keep progressively working on the research daily, achieving every task unit consistently until the actualization of the ultimate target of the research – creating new possibilities, getting the unsolvable solved, the uncreatable created, undiscoverable discovered, and the uninventable invented.

Every small unit of the research tasks accomplished and each marginal gain made in pursuing the ultimate research goal is a fundamental unit that contributes meaningfully to the overall achievement of the target invention. When you persistently work and complete all the tiny building blocks of the invention-making journey, the remarkable ultimate goal of the research gets achieved. The invention gets made excellently well.

The tiny task units of the research are the necessary building blocks of the entire investigation for making a remarkable invention. Every research task completed extraordinarily well, irrespective of how tiny, gets you closer by one step towards realizing the ultimate research goal.

Consistently get closer by one pace daily to achieve the overall research goal. In the long run, all the slices of the enormous research work get done excellently, and the utmost goal of the research gets astonishingly achieved.

At last, you get an essential invention, creation, discovery, or innovation made. Then, you victoriously exclaim:

- I have made an important invention.
- I've created an indispensable value.
- I've turned possible an impossibility.
- I've transformed unsolvable into solvable problems.
- I have impacted the world positively.
- I have created a highly demanded-global brand.
- I am known everywhere for my inventions.
- I have fulfilled my utmost human right to make inventions.
- I have changed and advanced the universe by one indelible giant stride.

CHAPTER 13

CHANGE IS GOLDEN FOR MAKING INVENTIONS

Change is golden for the making of inventions. Successful execution of research demands changes on the part of the investigator. A person must change how he typically does things to execute research exceptionally well and attain each research goal of value creation and invention making.
1. Change is necessary to make inventions.
2. Self-development to the fullest.
3. Self-development for making inventions.
4. Self-reexamination.
5. Focus.
6. Meditation.

CHANGE IS NECESSARY TO MAKE INVENTIONS

Make an uncomfortable change from what you are and how you do things to become a prominent researcher. Comfortable change does not make a person become an outstanding inventor.

With uncomfortable change alone, a person becomes an outstanding inventor. Only uncomfortable change leads to becoming an inventor. Uncomfortable change demands making unreserved sacrifices and a great investment of time, talent and treasure in the invention-making journey to guarantee excellent completion of the target invention.

Uncomfortable change results in better research goal actualization. The right change of behaviour leads to becoming a research goal-getter. Greater performance and new research goal attainment demand change in behaviour and the volume of research action exertion.

More uncomfortable actions exertion for making inventions culminate in better accomplishment of research goals. Substantially improved committed action is necessary for research completion. Investment of much more time, talent and treasure than what other people typically do in research leads to the making of valuable inventions and discoveries. Invest more of everything in and about you in making an important invention with your research.

Make an absolute determination to improve your research execution skills to the maximum of your potential. Put in the highest level of your resources, skills, talent and effort in carrying out research. Then, achieving the research goal becomes more certain for you.

Change your research execution performance to the rooftop of your potential. Consequently, each research goal you pursue becomes more bound to be attained. When done, making extremely valuable inventions and innovations shall become common results you get.

Make much more sacrifices than everyone else to get important inventions made. Unwavering determination and willing investment of whatever it takes are necessary to make exceptionally valuable inventions.

A person can only achieve change in his research execution by first changing his thinking. No one changes his research performance without first changing how he thinks and acts.

Change in thoughts is a prerequisite for a change in action exertion for the astonishing accomplishment of the invention-making journey. Change your thoughts. Change how you think in line with the demands of the inventions your research aims to achieve.

Think invention-making. Think discovery-making. Think creation of great value. Think about what, how, why, when, and where to make great innovations for the best good of humankind. Think exceptionally great invention-making thoughts.

Think and wholly invest everything in and about you in research execution to make indispensably indelible inventions, discoveries, creations, or innovations. Think research execution and act in a manner that consistently results in the timely making of the target invention, discovery or innovation.

Plus, embark on the research with a passionate commitment to solving the unsolvable and making the impossible possible. Execute the research thoroughly so that it discovers the undiscoverable, invents the uninventable, and creates the uncreatable.

What an individual expects greatly and works committedly hard enough toward actualization often gets realized by him. Aim at and have unflinching expectations of making the target invention in every one of your research works.

Execute the research so meticulously that what has been conclusively considered impossible because it is too good becomes possible. Carry out the research so rigorously that it consistently achieves the discovery, innovation, or invention that was set as the ultimate goal of the investigation.

Then, work dedicatedly toward its actualization. Keep exerting an ever-increasing right action on it until a thunderstruck accomplishment of the invention with your research.

SELF-DEVELOPMENT TO THE FULLEST

Develop yourself fully and make extremely valuable inventions to advance the world. Everyone has untapped talents, gifts, and unattained potential for value creation. Such talents or potential can best be reached and utilized only by a person's self-development. Put in all that it takes to fully develop yourself for research execution and invention making.

When an individual develops himself, he taps into his special latent gifts or talents and utilizes them successfully for research execution. Such research makes inventions that turn unsolvable problems into solvable ones and make impossibly good things to become possible.

By self-development, individuals change their thinking and actions. Then, they move to the next level of inventing solutions to societal and global problems. To make inventions, one must develop himself maximally.

Someone cannot invent products, solutions, ideas, or services that add great value to humankind without first developing himself. Self-development is one of the preconditions for making invaluable inventions and discoveries with research.

Seriously develop yourself to attain peak potential. Then, you can easily research and make world-changing discoveries and inventions with your research.

Fully develop your thinking, skills and actions for the most successful research execution. Do the research and make indispensable inventions and discoveries that change and perfect people, society and the world.

Execute research and hit one invention after another and one discovery after another. Research and create one highly valuable solution after another. Keep researching productively and become acclaimed globally as a prolific inventor.

Inventions alone progress the world the most. Inventions

emanate from research. Research is based on good ideas. Great ideas that change the world are the products of research based on radically good thinking. Revolutionizing thinking emanates from excellent self-development.

Therefore, the value of exceptional self-development in advancing the world through research is indispensable. Overwhelmingly, develop yourself for excellent execution of research to make inventions and discoveries that radically advance the world.

Develop yourself to think differently. Radically think differently to originate good ideas capable of skyrocketing the world's advancement. Thoroughly execute research that consistently makes the world's advancement a reality.

Make extremely valuable inventions and discoveries with your research to advance the world, improve society and improve people.

Exceptional self-development enables an investigator to fully focus on the ultimate research goal attainment while implementing the subgoals. An optimally self-developed person is always focused on the ultimate research goal achievement. Immediate challenges and distractions never enslave a fully self-developed person from the timely accomplishment of his invention-making journey.

Optimally self-develop and be completely prepared to get the most important thing for actualizing the ultimate research goal done at every moment. The most imperative task on which the attainment of the ultimate research goal depends should be done very well every single time.

Concentrate on doing the most crucial thing to accomplish your paramount research goal marvellously. The making of the invention and discovery that the research is aimed at should always come first and be fully attended to before everything else. Making solvable the unsolvable and turning possible the impossibly good thing with your research should remain your top priority and be accorded every attention to guarantee actualization.

None can invent anything of value without adequate self-development to embark on the research for inventing or discovering the thing. A person must self-develop to execute research before he can make an invention or a discovery of great value.

Exceptional self-development allows one to envision an indelible invention and exert unstoppable actions for its actualization. Nobody attains an invention he cannot visualize. A person does not take all the necessary actions to materialize an invention that is beyond his dreams. To invent, a person must be self-prepared not only to desire it but visualize the invention and do everything necessary to actualize the great dream.

Uncommon self-preparation is mandatory for the habituation of never giving up on an invention-making journey. Never ever give up on the research for making the target discovery, invention or innovation until it is accomplished.

Giving up is equivalent to living an unfulfilling life. An unfulfilling life is not worth living. Relentlessly press on and on until the astounding accomplishment of each of your research goals.

A person does not typically invent or discover a thing of immense value that he cannot see himself accomplishing. It is what an individual can see himself accomplish with a committed dedication that gets achieved when he practically demonstrates the level of dedicated commitment that the making of the invention demands.

Develop and prepare yourself fully for the making of an invention. Suppose you want to be the first and only person on earth to provide, invent, discover or create an indispensable solution with your research to a globally pressing problem. In that case, you must go all out in search of a wonderful solution to the internationally pressing problem before someone else does it.

To successfully get this done, the person must first change his thinking and work very hard and smartly to change himself from the inside out. To completely change oneself for the successful execution of research, the person must devour every research book on the problem under investigation.

Mastering and applying this book, RESEARCH, INVENT, AND CREATE WEALTH, and other research and statistics books by the current author are sure ways of marvellous self-development for making highly valuable inventions. The books include:

1. RESEARCH: MAKE IMPOSSIBILITY POSSIBLE: KPOLOVIE, PETER JAMES: 9798364274007: Amazon.com: Books
2. Amazon.com: MULTIVARIATE ANALYSIS OF VARIANCE: SPSS EXCELLENT GUIDE: 9798402243668: KPOLOVIE, PETER JAMES: Books
3. FACTOR ANALYSIS: EXCELLENT GUIDE WITH SPSS: KPOLOVIE, Peter James: 9798705490257: Amazon.com: Books
4. Amazon.com: CORRELATION, MULTIPLE REGRESSION AND THREE-WAY ANOVA: 9798595840255: KPOLOVIE, Peter James: Books
5. IBM SPSS STATISTICS EXCELLENT GUIDE: KPOLOVIE, Peter James: 9798563947115: Amazon.com: Books
6. Statistical Approaches in Excellent Research Methods: 9781482878301: Business Communication Books @ Amazon.com

7. Excellent Research Methods: Kpolovie, Peter James: 9781482824971: Amazon.com: Books
8. Handbook of Research on Enhancing Teacher Education with Advanced Instructional Technologies: Ololube, Nwachukwu Prince, Kpolovie, Peter James, Makewa, Lazarus Ndiku: 9781466681620: Amazon.com: Books

The books shall certainly make any user develop himself for exceptional research execution and invention making optimally. Study, master and apply the books to exceedingly self-develop, acquire, and apply new thinking skills and see everything differently as the making of each target invention demands.

Successful research execution, invention-making, problem-solving, wealth creation, income generation, discovery-making, opportunities creation and utilization, and value creation all depend on exceptional self-development, good thinking and action exertion. Exceptionally develop yourself. Then, and only then, you can research, make essential inventions or discoveries and enjoy all the glory and wealth that follow.

Extraordinarily develop yourself and radically improve your thinking tremendously for exceptional research execution. Dedicatedly exert the necessary actions in your research to bring good thinking to bear. Next, the inventions the research is aimed at shall be accomplished.

SELF-DEVELOPMENT FOR MAKING INVENTIONS

Self-development is indispensable for research execution and the making of inventions. A person must exceptionally develop himself to research and make essential inventions, discoveries, creations, or innovations to advance humanity.

With much more committed effort in action exertion, everyone who genuinely wants can develop, improve and grow towards attaining his maximum potential. It does not matter whether the person has the least or greatest talent in research execution. There are opportunities for everyone to develop himself and become better at inventing solutions to the persistent problems of humankind.

There are gaps in every person's invention-making ability that can only be filled when he committedly develops himself. Only when the person realizes that he has gaps in his abilities for making inventions will it occur to him to self-develop intensely and make serious improvements.

Realize that there are countless invention and discovery opportunities you are yet to develop yourself to successfully embark on the research to make the inventions. Engage in self-development and research to make more valuable discoveries, creations, and innovations.

Develop yourself to the peak at which you always focus completely on accomplishing tasks and activities at the moment. It is upon task completion at the moment that the attainment of your ultimate research goal of making inventions depends. Never back down on any of your invention-making journeys. Remain focused on exerting the right actions ceaselessly until the astounding completion of each of the target inventions.

SELF-REEXAMINATION

The making of inventions requires the investigator to reexamine himself from time to time. Self-reexamination is necessary for a person before he can become an eminent researcher and prolific inventor.

With self-reexamination, a person gains greater self-awareness and self-mastery in the science of research and invention-making. Develop yourself to attain greater confidence in your strengths and get a complete handle on your weaknesses.

Self-development increases a person's self-awareness. To self-develop, a person must closely reexamine himself. Reexamination of oneself increases a person's self-awareness. It optimizes the recognition of his talents, areas of strength and what to do to keep rapidly improving his areas of strength.

Self-reexamination also heightens a person's awareness of his limitations, inabilities or weaknesses. Then, it makes the person to master ways of overcoming his shortcomings.

An individual should work to maximize his areas of strength and minimize his weaknesses. Self-development is a function of reexamining oneself and taking every action to maximize his talents and eliminate or minimize his inabilities.

Self-reexamination increases a person's self-awareness and compels him to create abundant circumstances for his maximum potential accomplishment. Only when a person knows his best strength area that he can engage in doing everything it takes to maximally accelerate his strength.

Conversely, only by knowing one's ineptitude that the person can do all that it takes to get rid of it from slowing him down in his mission of attaining maximum invention-making potential. Thus, either way is necessary to reexamine oneself and do the needful for better research execution and more research goal attainment.

Without self-reexamination, there is no self-awareness or realization of one's weaknesses. Without self-realization, a person is dooming himself to failure in making inventions.

The people who remain in the dark about their strengths and shortcomings unwittingly doom themselves to failure in the making of inventions, innovations, and discoveries.

A person who is clueless about himself on the reasons he has yet to make any useful invention is never likely to make inventions of great value. Only when he becomes fully self-aware of his strengths and weaknesses that he can take every step to minimize or overcome the shortcomings and maximize his strengths in research and invention-making.

With the minimization of weaknesses and maximization of strengths, a person can research and make inventions of immense value. The individual has to overcome his limitations, inabilities and weaknesses to execute research successfully and make the target inventions.

To make inventions, the investigator must first know himself and apply himself completely as the making of inventions demands. Reexamine yourself. Know yourself to research successfully. Plus, know yourself to make important inventions or discoveries.

The person unaware of his shortcomings in research execution and invention-making cannot overcome them. He cannot successfully conduct research and make the target invention without overcoming his shortcomings. Perhaps the only two things hindering him from making essential inventions are:

1. Not dedicatedly maximizing his areas of strength in research execution.
2. Not committed to minimizing his areas of weaknesses in research execution.

The invention-making demands a person to self-reexamine and know his limitations and strengths. Plus, overcome the restraints by best minimizing them. Capitalize on his areas of strength by best-doing everything to maximize them.

To know one's shortcomings in research execution and invention-making, the person could learn and practice the execution of research much more. He could teach others research execution skills, steps and aspects for successful invention making.

Suppose he needs help teaching research and invention-making skills, steps and processes. Then, he has yet to grasp the matter, skills or information completely. Therefore, the person must self-develop and attain expertise in the skills for research and invention-making.

When the person does not reexamine himself, he does not realize his weaknesses. When he does not know his shortcomings, he makes little or no serious effort to learn, practice, and master the missing research execution skills.

Failure or refusal to self-examine, know one's weaknesses and take actions to overcome them results in a two-fold setback that could be referred to as a double whammy. Only with a double whammy will a person convince himself that he has the research execution skills he does not have.

Only when the person overcomes the double whammy that he can deliberately, persistently, excessively, repeatedly, and passionately practice over and over, again and again, to attain super-mastery of the invention-making skills. When he practices and attains super-expertise in research execution skills, he can easily research and make extremely valuable inventions outstandingly.

Therefore, reexamine yourself and know your strengths and weaknesses. Do everything necessary to maximize your strengths and overcome your weaknesses in research execution. Only then can you easily accomplish your invention-making journeys victoriously each time.

FOCUS

Focus completely at each time on the accomplishment of the experiment upon which making the target invention of your research depends. Focus intensely on completing every detail of the experiment essential for timely achieving the target invention of your research.

Train your mind. Take full control of your mind, attention and actions to conclusively pursue the attainment of one research goal before another. Concentrate everything on the actualization of one research goal before embarking on another. Withdraw from everything else to concentrate all attention on the successful completion of each invention-making journey you embark upon.

Tune out every one of the distractions that have prevented other people from making the invention your research targets. Do whatever it takes to achieve the invention your research targets astoundingly.

The concentration of efforts and resources on accomplishing each research journey is necessary for making the invention that the research aims at. With intense, long enough practice, a person can improve his focalization to pursue the making of an invention to a successful completion.

At each time, please focus on the best exertion of actions on the most important task at hand for the dazzling actualization of the invention or discovery the research aims at until it is awesomely done. Triumph in making the invention the research targets is a function of focus and single-minded concentration of all mental and physical actions and resources on completing the invention.

Most often, making an invention or discovery demands absolute attention on the experiment upon which the invention making depends. Absolute attention can only be accorded to an experiment or research when we completely focus or concentrate our time, talent and treasure on making the target discovery and creating value.

To make an invention of incredible value with our research, we must be tuned into what we must do and do only that until we attain the ultimate research goal. The making of an invention is a task that demands the ability to block out all unnecessary and unwanted distractions, information and obstacles to concentrate on one thing: the research on which the invention depends.

Focused attention has largely become the casualty of our high-tech age. Each person now contends with an astonishing volume of distractions in this highly information and communication technology-led age than ever before (Kpolovie and Lale, 2017).

Attention lapse has become a more common experience than in the past. As a result, the making of the invention now demands far greater training of the mind to overcome all distractions, prevent all interruptions, and defeat all multi-tasking temptations. The making of invention demands concerted training and mastery of our minds to focus, concentrate, single-task, and accord undivided attention to the experiment or research upon which the invention-making depends.

The human body has approximately 100 billion neurons, of which roughly 86 billion are in the brain. The neurons function like super-high-speed neural processors. Each adult human also has about 85 billion nonneuronal cells.

When the neural and nonneuronal cells are concentrated on executing the most deciding research execution task each time, the likelihood of successfully making the target invention, creation, innovation, or discovery is optimized. That is why each researcher is charged to focus entirely on the execution of the research for the marvellous making of the target invention.

The likelihood of successfully making an exceptionally valuable invention, discovery, innovation, or creation with our research increases to the fullest when we:
1. Focus all the billions of neurons that function as super-high-speed neural processors on making the invention.
2. Dedicate all the billions of nonneuronal cells to making the target invention.

3. Concentrate all our time on executing the research upon which making the target invention depends.
4. Invest all our talent, treasure and resources in the phenomenal completion of the experiment for making the target invention.
5. Exert all our efforts and take far more than all the actions the invention-making demands.

Therefore, completely focus all the billions of your neurons on executing the research to make your target invention. Singleheartedly dedicate the billions of your nonneuronal calls to making the target invention of your research. Completely concentrate your time on executing the experiment for making the invention the research is aimed at.

Unreservedly invest your treasure, talent, and resources in the phenomenal completion of the experiment upon which making the target invention depends. Exert every effort, make every sacrifice, and take far more than all the actions that exceptionally completing the research for making the invention demands. Consequently, the invention more certainly gets made.

Never let any room for distractions that will drift your attention and actions away from working nonstop on making the target invention. Consistently push on and on with an ever-increasing momentum on making the invention your research aims at until it is fabulously achieved.

Brain-train to single-task on the research or experiment upon which the target invention lies. Engage in meditation or yoga methods that best reduce distractions while simultaneously maximizing action-backed focus on making the invention your research ultimately aims at.

MEDITATION

Practice meditation training techniques that could make your brain most efficient in concentrating its entire billions of neural resources on the execution of the research upon which making the target invention hangs until the superb attainment of the goal.

Meditation sharpens focus on the goal and improves the concentration of actions on the research for the timely outstanding accomplishment of the target invention, creation, discovery, or innovation. About 5 to 10 minutes of daily meditation on making the invention, which is the ultimate goal of the research, could result in a more efficient concentration of efforts towards the exquisite actualization of the ultimate research goal.

Daily meditation practice could lead to a better focus on the research execution. The improved focus could result in enhanced learning that, in turn, advances the execution of the research for the best attainment of the invention.

Meditate for 5 to 10 minutes daily on making the invention that is the ultimate goal of your research. Meditation optimizes the ability to concentrate on the research execution. Meditation increases the brain's power to actively focus single-mindedly on making the invention, discovery, or innovation the research targets.

With meditation, the brain is trained to improve its level of attentional control for the best accomplishment of each invention-making journey. Greater attentional control leads to more timely exceptional accomplishment of each research journey.

Meditation helps in the most effective and efficient prioritization and completion of the different research tasks, units, or subgoals until the attainment of the ultimate goal. Even a small dose of meditation daily could improve concentration, focus and attention on working practically to best accomplish the research for making the craved invention.

The radically changing world we live in demands a person to change himself suitably before he can invent a product of great value. One must self-develop optimally to invent or create value capable of advancing the world. A person must reexamine himself and exceptionally improve his areas of strength for superb completion of the invention-making journey he embarks on. A person focused on making inventions has to keep developing himself for excellent execution of the invention-making research. Please do everything the exceptional making of the target invention demands.

CHAPTER 14

USE EVERY OPPORTUNITY TO INVENT

Every person who truly wants can become a profound inventor. Each renowned inventor worked very hard and smartly in self-development, utilizing opportunities to intensely research and make life-changing inventions. You, too, can do so and become a prolific inventor.
1. Utilization of every opportunity to invent.
2. Create opportunities for making inventions.
3. Think good and invention thoughts.
4. Meditation and discontentment creativity.

UTILIZATION OF EVERY OPPORTUNITY TO INVENT

Never has anyone been born a profound researcher without working tirelessly to become one. Capitalize on the abundance of opportunities to research and make essential inventions.

Every problem is an opportunity for research and the making of inventions. Every need of humankind and each of human's wants is a ripe opportunity for research execution to meet the need or want permanently.

Opportunities for research and invention-making usually come wrapped in problems, needs or wants. Each research and invention-making opportunity is an idea everyone else has simply missed. Or, an idea which none else has imagined and nobody has successfully worked on for exceptionally solving a problem, and meeting a need or want of humankind.

View every problem, need or want with a fresh eye to see and utilize the opportunity for invention, creation, innovation, or discovery-making concealed in it. Next, embark on the research for remarkable accomplishment of the invention, and get the great need met.

Take every risk that the making of the invention demands. Make all the sacrifices in researching to accomplish the target invention or discovery. Every phenomenal researcher or inventor takes the greatest risks and makes untold sacrifices that other people have not made and would not dare to make.

Take substantial risks that you have a personal stake in the outcome. To make an invention, you cannot sit on the sidelines playing it safe or doing only the things others have successfully done.

Those who prefer to play it safe and the people who are delighted in doing only things that another person has successfully done never make inventions, creations, or discoveries of any kind. What another has done, created, invented or discovered, somebody else cannot do in the same way and become an inventor, discoverer or creator of that thing.

Become a go-getter. Take the greatest amount of initiative to set the making of invaluable inventions as your research goals. Dedicatedly invest much more than all the necessary actions unwaveringly to translate the goals into reality.

Remain locked on to doing the arduous tasks that others have yet to do as long as it takes to achieve the set research goals. Intentionally and deliberately take all the steps, no matter what, to remarkably accomplish the ultimate research goal in each invention or discovery journey.

USE EVERY OPPORTUNITY TO INVENT

Let making inventions with your research be the most important thing you want out of your life. Never allow the mountain of distractions, stressors and costs to interfere or prevent you from attaining that life goal. Do whatever it takes to outstandingly accomplish the inventions, creations, discoveries, or innovations your research aims at.

Take all the actions to utilize the opportunities for the making of inventions. With the exertion of the right research actions, inventions are made. Take much more than the necessary actions to translate the invention-making opportunities around you into reality. Outstandingly execute the research for making those inventions.

Move from intention into action and keep exerting ever-increasing action in executing the research until the astounding achievement of the target invention in each research journey. At all times, let the making of a phenomenal invention fuel and motivate you to increasingly accelerate your effort to overcome all obstacles and distractions until the wonderstruck accomplishment of the life-changing invention your research aims at.

Persistently take all the steps and exert all the actions from today steadily, reliably and consistently until the attainment of the ultimate goal of every research you embark upon. Never put off any of the things you should do today for tomorrow or another day. Do the research from today for the best actualization of the set goal of invention-making.

Action, irrepressible action in research execution, moves us to the actualization of our invention-making destiny. You have the potential of inventing, creating or discovering solutions to certain major problems of humankind. Maximize utilizing that potential.

However, without executing or funding research for making those discoveries, inventions or creations; the products, solutions, services, or ideas cannot be materialized or achieved. You will not even realize that you have the abilities within you to make such inventions when you do not embark on or sponsor the requisite research.

The creative talents within a person cannot develop themselves for the person to achieve his destiny. A person's life mission of making essential inventions can only be achieved when he takes inextinguishable actions in executing the research on which the attainment of the inventions depends. Enthusiastically embark on research for making inventions, discoveries and innovations that greatly improve people, develop society and advance the world.

CREATE OPPORTUNITIES FOR MAKING INVENTIONS

Create invention-making opportunities and utilize them to the fullest. Never wait to seize the opportunity. Instead of waiting, create opportunities in everything, every circumstance and all the time to exert peak actions to attain each of your ultimate research goals.

Create and fully use opportunities in everything and every condition to maximally advance towards accomplishing the inventions and discoveries your research aims at. Opportunities for exceptional actualization of your inventions and discoveries should always be created and utilized maximally by you.

Then, let the accomplishment of your inventions and discoveries create unlimited opportunities for other people to seize to advance toward attaining their apex potential. Be the pathfinder. Your inventions and discoveries should create the path for other people to better accelerate toward attaining their maximum potential.

Each of your accomplished research should serve as a bridge on which people chart uncharted territories and tread new horizons. Every one of your inventions, innovations or discoveries should be more indispensable and vital than the unshakable foundation upon which people build countless structures for global advancement.

Let each of your inventions be the solid rock on which the advancement of the world depends. Your findings should compel others to research, accelerate to the apex of their potential, and advance the world astronomically.

THINK INVENTION THOUGHTS

Think good thoughts. Always think invention-making thoughts. Think discovery thoughts. Think creation thoughts.

Daily think innovation thoughts. Originate invention thoughts. Initiate invention ideas and take all the actions for their materialization.

Always think and fashion novel ways of making marvellously good inventions. Only when you committedly think this way that having invention, creation, discovery and innovation epiphanies shall become your dominant habit.

When an excellent invention-making idea intrigues you, please write it down. Keep it ever before you. Take every necessary action, no matter what, to bring the invention idea to the fullest materialization or actualization in this physical world.

Let great thinkers, productive researchers, and prolific inventors be the closest people in your life. Let them be the people with whom you spend quality time. Read about them, their works and inventions. They are the right people to socialize with and spend time with.

The right people sharpen one another with their exceptionally good thoughts of invention, creation, discovery and innovation ideas. Spend much more engaged time with such people. Then you will have a smoother journey to making solvable unsolvable problems and possible impossibly good things.

Always think superb invention thoughts, and great innovation thoughts. Generate excellent visions that will add much value to people's lives, perfect society and advance the world. Take dedicated actions on the thoughts and ideas with your research until actualizing the envisioned inventions, discoveries or creations.

Think great invention thoughts. With total commitment, exert unreserved actions always to actualize the inventions.

Good invention thoughts, great discovery ideas, and committed exertion of extreme actions culminate in making highly valuable inventions, superb products, marvellous discoveries, and the rendering of wonderful services. The type of inventions, products, innovations, discoveries and services that will radically improve humankind and make each person accelerate swiftly toward attainment of their maximum potential and make the world a much better place for habitation depend on the quality of thoughts generated by the researcher.

To keep achieving success from glory to greater glory with your research, continue to develop and improve yourself. Self-improvement demands continuously reading, thinking, meditating, writing, practising, learning, and, most importantly, experimenting. Aptly, self-development demands one to research.

Grow yourself by filling your mind with good thoughts, thoughts of making great inventions and discoveries to improve people, perfect society and advance the world. Put the great invention-making thoughts into action.

Keep exerting the right actions upon the good thoughts to make impossibly good things possible from time to time. Take all the necessary actions to make the great inventions, innovations and discoveries that preoccupy your mind for the advancement of the world. Execute exceptional research for the materialization of good thoughts.

Write and publish the results of your research as science books to enable countless people to acquire them and benefit from your research. The more people acquire and apply the books, the better society and the world can advance.

Think countless great invention thoughts. Research on them. Make the inventions and discoveries that actualize the great invention thoughts.

The inventions will propel countless others to crave invention-making empirically. Consequently, people will accelerate faster toward attaining their maximum potential. Then, the world shall become a much better place to dwell.

Good thinking, acted upon with unwavering research, produces good inventions, products, ideas, services, discoveries and innovations that change life for the best. Research successfully done based on great thoughts culminates in great inventions and valuable products.

Always learn not only to be a great thinker but someone who meticulously brings every one of the great thoughts into reality via intense action in research execution. Please research to actualize each of your noble, good, great, and astronomical invention-making thoughts.

The person the world needs the most for its sustainability and advancement is someone who thinks good inventive thoughts always and takes all the necessary actions by meticulously researching to actualize the great thoughts. Become that most needed person to better people, society and the world.

Let the products of your research shine so brightly to illuminate the world permanently. Your research output should positively impact the advancement of people, society and the planet.

Habitually originate great invention and discovery thoughts. Take rigorous actions with your research to materialize your great thoughts. Next, implement your thoughts with your research. Then, the improvement of people, the betterment of society and the advancement of the world will automatically follow.

MEDITATION AND CONSTRUCTIVE DISCONTENTMENT CREATIVITY

Dedicated meditation is practised daily by prolific inventors to generate and bring their invention-making ideas to fruition. To most successfully execute research, solemnly meditate, think and activate your mind's full creative, innovative and inventive potential.

Practice meditation and constructive discontentment creativity. A constructive discontentment creative person is not only dissatisfied with things as they are but committed to executing research passionately for practically turning them into the best that can be imagined. Each time, be successfully engaged in research to turn something into the best it could be.

Self-evolution to become an eminent researcher or prolific inventor demands a lot of dedicated meditation via the beta through alpha to theta level (Sigmund Freud, James Strachey, and Peter Gay, 1989). Meditation allows for both conscious and subconscious generation of ideas and methods of executing research to arrive at invention of immeasurable worth to advance people, society and the world.

Infinite awakening of mental alertness is achieved at the beta level of consciousness meditation. The optimal learning, growth and self-awareness state for solutions generation is reached with the alpha meditation level. With a blend of meditation and deep sleep, enhanced introspection, self-hypnosis and body-mind healing, a much more incredible invention-making is accomplished at the theta meditation level (Ed Bernd Jr, 2000; Jose Silva and Philip Miele, 2022).

The more inventive, innovative and discovery ideas, thoughts and insights a person generates via daily meditation, the more creative he becomes. As a person becomes more creative, he gets better at successful research execution. When the person becomes more successful in conducting research, the greater and more valuable impossibilities and unsolvable problems he makes possible and solvable.

In a lifetime, no person has been able to exhaust his creativity, and perhaps none could. Creativity is inexhaustible. Therefore, daily generate creative ideas, opinions, options and thoughts for successful research execution to best meet humankind's needs.

Again, the more creative a person becomes, the closer he gets to making incredible inventions, discoveries or innovations with his research. Individuals can only invent, discover, innovate, and turn impossibilities into possibilities and unsolvable problems into solvable ones by being exceedingly creative.

A person can only invent something with his research if he is much more creative than all the others regarding the target invention in question. A person cannot be more successful in research execution and invention-making than the magnitude of his demonstrated constructive creativity.

Seek new, revolutionizing ideas for solution discovery all the time. Invention, discovery or innovative research can only be done by a highly creative person. A highly creative individual typically seeks new, interruptive ideas to better advance the world.

Being exceedingly creative humbles a person and makes him much more teachable. Creativity makes a person become a better learner. A more teachable person and a better learner alone can master the science of research execution.

Successful research execution, making the unsolvable solvable and the impossible possible, depends largely upon constructive creativity. Making inventions, innovations, and discoveries depends on constructive discontentment and creativity. Become an exceedingly creative thinker by daily generating inventive and innovative solution thoughts via meditation, practice, and experimentation.

Only with constructive creativity can one research, invent, innovate and discover things of great worth. With exceptional constructive discontentment creativity alone, can a person make solvable the unsolvable, and turn possible the impossibility via meticulous research.

Constructive discontentment creativity is needed much more than everything else for inventive research execution.

As stated earlier, a constructive discontentment creative person is not only dissatisfied with things as they are but committedly executes research passionately for practically turning them into the best that can be imagined. Each time, be successfully engaged in research to turn something into the best it could be.

An uncreative person spends years, often an entire lifetime, without identifying any researchable problem. When he cannot even identify a problem, is it functional solutions that the person can originate? No. Certainly no.

An uncreative person makes no attempts to search empirically for a permanent solution to humankind's pressing, persistent problem. For the uncreative person, nothing is a problem that should permanently be solved.

Always overwhelmingly demonstrate constructive creativity. Get exceptionally creative so you can successfully embark on or fund research execution.

Please, become exceedingly creative. Only then can you research and attain the status of an originator and generator of empirically workable ideas and solutions. Only then that you can turn impossibilities into possibilities and unsolvable problems into solvable ones for the best advancement of society, humankind and the world.

Always be creatively engaged in research execution actions and tremendously add value to people. Always do research-related works that add much more value to people's lives.

Only engagement in or sponsoring creative and inventive research execution adds value to people. Research and invent solutions to people's pressing problems.

Cultivate constructive discontentment creativity. Nurse it by passionate practice till you become a constructive, creative giant. Through practice, become a cauldron of intense discontentment and constructive creativity brought into reality with your experiments.

Work very hard and smart regularly to become a typical creative thinker. Meditate, think outside the 'confining lines', 'limiting rules', and 'conventional constraints' to alter the status quo for the best.

Characteristically research intensely to invent greatly valuable products, solutions and services. Research and invent to become distinguished as the all-time next game-changer in the world.

Think extraordinarily creative all the time. Originate workable ideas until you popularly become the voice of reason, solution generator, and unsolvable problem solver. Research, create, invent, discover, innovate, and turn impossibilities into possibilities. Keep inventing indispensably valuable things with your research until you become globally renowned as one of the most prolific inventors on the planet.

CHAPTER 15

TAKE EXTREME ACTIONS TO INVENT

Every invention is a product of exerted extreme actions. Taking extreme actions must always come before an invention or value creation is accomplished. Invention-making demands taking extreme actions. Extreme actions must always be taken before an invention-making journey gets completed. Take extreme actions to invent, create, or discover value the world badly needs.
1. Extreme actions result in inventions.
2. Be different from all else to invent.
3. Triumphantly overcome all the insuperable.
4. Be fanatical and obsessive about inventing.
5. Be extraordinary to invent.
6. Take total responsibility for inventing.

EXTREME ACTIONS RESULT IN INVENTIONS

Extreme action is the type of action that all others do not take because it is too extraordinary. Extreme action is the type of action that sets a person distinctly apart from all the rest people. Extreme action is the sort of action that makes the person who repeatedly or regularly takes it to attain dominance in his profession and in the global marketplace. Taking extreme action is doing the exceptional things that others have not done, cannot do, refuse to do, and will never be the first to do for the making of the invention in view.

Every great thing happens with action. Without action, nothing good happens. For everything, there is a cause. The extreme actions that a person takes cause invention. Somebody can only make an invention when he takes all the extreme actions it demands. To create value or make an invention, take all the necessary extreme actions.

Every invention, discovery, creation, and innovation can only be made by taking the extreme actions it demands. It takes the taking of extreme actions for an invention to be made. To invent, one must take concerted extreme actions and excessively do everything that making the invention demands.

A person does not make an invention without taking extreme actions. To make an invention and create value, a person must take all the necessary extreme actions. Take extreme actions for the excellent making of your target inventions.

With extreme actions, a person, team, company, organization, multinational corporation, state or nation gets an invention, discovery, creation, or innovation of immense value made for humankind and the world's advancement. By doing only the things people do, to the extent they do, a person, team, company, organization, multinational corporation, state or nation cannot invent anything.

If by doing things the way people do can get a particular invention made, the invention could have long been made by them. The reason why the invention your research aims at making has not been made by others is because there are certain magnitudes of actions that nobody else has taken for that invention to be made.

Those extreme actions that others have not taken are the reasons why the invention you aim at making has not been made by anyone else before you. Only when you deviate from what others have done and take extreme actions that have never been taken before that the target invention gets actualized. Without any hesitation, please take those extreme actions and get the invention in view made extraordinarily well.

Remain totally committed to taking extreme actions for the attainment of each goal your research aims at. Only those who passionately, persistently, and committedly take enough extreme actions achieve the inventions their research works target. Keep taking passionate, extraordinary actions to actualize each invention your research targets until you become renowned as a dominant, prolific inventor, creator, discoverer, and innovator in your chosen profession.

There is no number of inventions a person, collaborating team, company, organization, multinational corporation, state, or nation makes that would be too much. The more inventions you make, the better people, society and the world become. Each invention advances the world by one giant leap.

In like manner, there cannot be any number of extreme actions a person, collaborating team, company, organization, multinational corporation, state or nation takes to actualize her invention dreams that would be too much. The more extreme actions are taken for invention-making, the better. Every extreme action exerted to make an invention gets the investigator closer to accomplishing the target invention.

The great news is that regardless of a person's enormous extreme actions to make inventions, he can never run out of his creative and inventive talent. A person cannot run out of his potential to make inventions, discoveries, creations, and innovations. A person's energy for invention-making cannot be exhausted.

The more the magnitude of talent a person utilizes for making inventions, the greater he becomes much more talented for value creation. The person who has originated 100 inventive and creative ideas and persistently pursued and actualized each of them with passion will always have much more creative and inventive ideas than the person who has originated only one idea and pursued it to materialization.

No person has ever exhausted his creative and inventive potential. A person must always have inventive ideas, energy and efforts, regardless of the magnitude of extreme actions he has taken to get inventions made. The more inventions a person makes, the greater his talent, potential, energy, resources, and efforts to exert much more extreme actions and make far more worthwhile inventions.

Each extreme action you exert to make an invention radically sharpens, strengthens, and thoroughly increases the capacity of your talent, potential, and experiences to make much more valuable inventions. Contrary to reducing, every extreme action a person exerts for making an invention extensively and utterly increases his talent, potential and experience for the exertion of far more extreme actions that result in the accomplishment of much more useful inventions to best meet the needs of humankind.

Consequently, making one invention propels, catalyzes, motivates, and enables the researcher to embark on more demanding research for triumphantly making much more essential inventions. When a person excellently makes an invention, he gets much more enabled and charged to compulsively take greater extreme actions for the astounding making of more extremely valuable inventions.

Please, exert extreme actions to get your target invention made. Set a more demanding invention-making goal. Exert the needful extreme actions and fantastically accomplish the invention. Keep replicating the process for the actualization of much more inventions.

BE DIFFERENT FROM ALL ELSE TO INVENT

Nature only lets a person make an invention once he overwhelmingly improves himself, his thinking, actions, and approach to the target invention. It is only when a person exceedingly fortifies himself and consistently, persistently, resiliently, and passionately invests his unreserved potential, talent, energy, treasure, time, and action in the invention-making journey in an exceptionally greater than the levels that no one else is willing to do that he can get the invention made. When a person sacrifices everything about him overwhelmingly more than anybody else has done and no other person can do to make a target invention, nature lets him accomplish the invention.

One must clearly distinguish himself from everyone else by investing his resources, creativity, effort, and energy in making a particular invention until it gets done. Only the person who invests everything in and about him in the invention-making journey attains the desired goal – making the invention.

Doing exactly what others before you have done or what any other person is doing to make a particular invention cannot enable you to become the inventor of the target product. To invent a product, you must do what no other person has done; none else is doing, and nobody else can be the first to get it done.

A person must do what others have not done, will not do, and cannot be the first to do for making the target invention. He must originate a unique vision, the type that no one else can be the first to have and pursue to outstanding completion. He has to necessarily go the length where others have not, will not, and cannot go in order to get the target invention made.

The person must exert the volume of actions the world considers unreasonably extreme in the invention-making journey. Only then can he get the target invention made. It is when you keep doing those things that only you would and could

TAKE EXTREME ACTIONS TO INVENT

do, those extreme things that only you are and can be willing to do, that will enable you to make the invention in view. Please, persistently do what nobody else can do to get highly useful inventions made.

The world does not know you and will never know you until you make that invention of your dream exceptionally well. Make that essential invention. Create that dominant value in the world. The value can only be created when you think and do what none else has done; nobody else is willing to do it, and no person other than you can be the first to accomplish it on the planet.

TRIUMPHANTLY OVERCOME ALL THE INSUPERABLE

Let the goals you set for your research be as overwhelming as, if not greater than, the ones set by the giants on this planet who have implemented their plans and accomplished the goals, made the target inventions, and created massive wealth. Set, pursue and accomplish goals greater than or as extreme as those by the Bill Gates and Paul Allen (founders of Microsoft), Steve Jobs (founder of Apple, NeXT and Paxar), Larry Page and Sergey Brin (originators of Google), Brain Action and Jan Koum (WhatsApp founders), Martin Eberhard, Marc Tarpenning and Elon Musk (founders of Tesla Motors that manufactures Electric Vehicles), Henry Ford (Ford Motor Company founder), Zhang Yiming (founder of TikTok), and Jeff Bezos (founder of Amazon.com).

Invent, create, discover, and innovate products that perpetually change our planet with your research. Embark on research aimed at inventing products that forever change the world for the best. Invest everything in and about you in executing the research until the actualization of the target invention.

With such a substantially overriding goal as your research's target, no resistance, obstacle, or challenge can be huge enough to prevent you from investing everything to achieve the invention. When the ultimate goal of your research is greater than everything else, you can surely surmount every seeming insuperable difficulty that comes between you and the goal. Then, you awesomely accomplish the target invention.

Regardless of the number of seemingly unsurmountable obstacles that come between you and your ultimate research goal, victoriously conquer each of them and get the target invention fabulously made. Surmounting every one of the obstacles is key to the making of an invention. Everything really worth doing in an invention-making journey is and should be done several times, repeatedly over and over again, until the target invention is astoundingly actualized.

In life, generally, anything worth doing deserves to be done many times, repeatedly. Overcoming every obstacle in the pursuit of an invention is worth doing, and it must be done as many times as the difficulties and challenges that tend to prevent the researcher from completing the invention. Triumphantly surmount all seeming indomitable difficulties, regardless of how many they may be, to accomplish the target invention marvellously.

BE FANATICAL AND OBSESSIVE ABOUT INVENTING

Obsession is a natural, creative and inventive human state that keeps a person completely fixated to extremely doing everything needed toward the actualization of his ultimate research goal until the target invention is excellently accomplished. It is the intense interest in and innate commitment to be completely engaged in extreme exertion of actions to accomplish a person's invention-making goal until it is awesomely achieved.

Obsession or being fanatical is a person's undying desire and total commitment to exerting unreserved actions toward the excellent accomplishment of his ultimate research goal.

Be fanatical and obsessive about making an invention. The making of an invention demands pursuing it fanatically. A person must exhibit extreme enthusiasm and total devotion in extensively exerting every action for making his target invention before it can be accomplished. Be fanatical about making an invention and exert extreme actions in the pursuit of the invention before it can be achieved.

Throughout the history of invention making, each inventor exhibited fanaticism for the invention he set out to make until it was achieved. Each invention on Earth was pursued fanatically and obsessively to completion by the inventor. Every invention was accomplished with extreme actions. For an invention to be made, the researcher must be fanatical and obsessive about it in all his approaches.

Develop and sustain obsession for the making of every invention your research targets. Be obsessed with making the target invention and invest everything into its pursuit until it is made exceptionally well. With obsession, massively exert every action required for superbly making the invention.

Demonstrate a compulsive urge and craving for making the invention, innovation, discovery, or creation your research aims at. Get a frenzy about making an invention for it to be actualized.

Always exhibit an intense need to quickly complete the target invention of your research. Then, exert unreserved actions toward its actualization, nonstop, until it gets wonderfully accomplished.

When a person is obsessed with achieving his target invention, he will not compromise doing anything necessary to make the invention. Only when a person is fanatically and obsessively serious about making the invention his research aims at that nature let him achieve the goal.

TAKE EXTREME ACTIONS TO INVENT

Nature is very reluctant to reveal the solution to an unsolvable problem. Only when you are frenzy about making an invention that nature let you make it. You must demonstrate 100 per cent commitment, complete and utter conviction that you will persistently pursue making a particular invention until it is made, no matter what, before nature will let you make the invention.

Then, nature smiles at you and unveils the target invention for you. For every invention made, the researcher demonstrated an untold commitment, obsession, and fanaticism before nature allowed him to achieve the goal.

Please demonstrate an obsessive preoccupation with doing everything to get the target invention made. An uncommon persistent concentration of actions on getting an invention made is mandatory for a person to invent the product he targets.

Pursue the exceptional attainment of your ultimate research goal, your target invention, with undying and persistent obsession as though your entire life depends upon its actualization. Work each hour of every day with an obsessive focus on making the invention your research targets. With utmost commitment, passion, gusto, and grit, exert extreme actions toward actualizing your target invention until it is superbly made.

To achieve the invention a research work targets, the investigator must be obsessively consumed by the lofty goal and work committedly nonstop every hour of each day on actualizing it until a thunderstruck completion of the invention. A researcher must be fanatical about the actualization of his invention-making dream before it can be achieved.

He must obsessively take all the extreme actions necessary to timely get the invention made. Be obsessed, fanatical, fixated, and get a frenzy about passionately doing everything, no matter what, for the timely excellent materialization of your invention-making dream.

TAKE EXTREME ACTIONS TO INVENT

Have irrepressible compulsion for fantastic completion of your target invention. Get addicted to pursuing your target invention until it is exquisitely achieved. Persistently demonstrate an intense interest in and fascination for the actualization of your target invention until it is marvellously completed.

BE EXTRAORDINARY TO INVENT

Never settle for the ordinary if you are to amazingly accomplish the making of an invention. The ordinary never leads to any value creation. Always think extraordinary thoughts to visualize what to invent and the value to create. Set extraordinary goals with your research. Exert extraordinary actions to achieve the inventions you aim to make.

Execute the research extraordinarily. Plus, always attain thunderstruck accomplishment of the ultimate research goal in each invention-making journey. Make extraordinarily useful inventions, discoveries, creations, and innovations to advance the world.

Extreme actions alone culminate in extraordinarily useful inventions. Exert massive actions in your research endeavours. Attain extraordinary success in every research you execute.

Extraordinarily enormous success in making inventions demands complete investment of everything in and about you in executing the research. Invest everything you have and are in the making of the target invention. Make a complete investment of your resources – your time, treasure and talent, far beyond what all others have done unsuccessfully in an attempt at making the target invention. Then, the invention aimed at shall get extraordinarily made.

Invest incomparably more resources, time, talent, energy,

effort, skills, and creativity in your research execution journey than every other human has unsuccessfully done in pursuit of the invention your research aims at. Only then can you attain exceedingly useful results than anybody else has done or could do. Only then that your target invention can be exceptionally made.

TAKE TOTAL RESPONSIBILITY FOR INVENTING

Take complete responsibility for making your invention. Making an invention happens only when a person, collaborative team, company, multinational corporation, state, or nation responsibly pursues it to fruition. You must take full responsibility and extremely invest in it for the target invention to be made. Please, practically demonstrate that utter responsibility. The only source, generator, originator, and architect for an invention to be made is the inventor.

You are the person solely responsible for an invention to be made. Singleheartedly take full responsibility for inventing something of great value to humankind. Committedly fulfil your unique right, responsibility, obligation, and duty to humankind by making an invention of tremendous worth. Make significant breakthroughs in your chosen field by creating or inventing crucially valuable products, solutions, and services that increase your dominance in the marketplace, incredibly improve people's lives, and skyrocket the world's advancement.

CHAPTER 16

CREATE THE FUTURE WE CRAVE

With research, everything good in the past was invented. All the desirable things at present were invented with research. We can invent every one of the great things we crave with enough rigorous research.

1. Actualize the future we crave.
2. Only the audacious makes inventions.
3. Dogged perseverance in invention making.
4. Exert the infinitely creative power within you.
5. The greatest investment to make.
6. Become the light that illuminates others.
7. Seek synergy.

ACTUALIZE THE FUTURE WE CRAVE

It is with research alone that we can invent or create the blissful future of our dreams. We should keep embarking on research to achieve our greatest goals.

Invention-making with our research today is superior to everything else in our quest for a blissful future. Success, happiness, peace of mind, wealth, a better life, and all the good things we crave depend upon the research we currently do. Only with research can we make discoveries and inventions of tremendous value to best meet humankind's great needs.

The master key to genuine success and achievement tomorrow is our research today. We can attain true success and lasting achievement tomorrow only with our research today that invents, discovers, innovates, or creates invaluable devices, products, services, ideas, solutions, and knowledge.

Just as the past and the present were structured by research, what becomes the future of humankind depends unilaterally on research. The research we do today alone that best determines our tomorrow, our future.

Suppose truly we are to have a better future. In that case, we must execute all the necessary research today to materialize the desired tomorrow. Our tomorrow is a direct function of our research today. Today, we get to continue researching and inventing the tomorrow of our dreams.

No man is happy who does not think of having a better tomorrow. Thoughts and hopes of having a better future give us happiness. Suppose that feeling of happiness is not to be a sheer delusion. In that case, we must research today and make all the requisite inventions to materialize it. A better tomorrow can only be attained with the quality and quantum of inventions and discoveries we make with our current research.

We have deeply within us the divine creative potential to execute research and make all the inventions upon which a better future depends. We must, therefore, utilize the infinitely creative spark within us to execute research for the inventions of a thriving future.

An enraptured tomorrow depends on the valuable inventions and discoveries we make at the present with our research. There is overwhelming evidence from all perspectives that success and advancement of the world are a function of the inventions we make today with our research than with all other factors put together. Therefore, we must research today and make all the needful discoveries and inventions that the better tomorrow demands.

Research execution and invention-making constitute a primary principle for achieving real success in life. A person who researches and makes valuable inventions enjoys fortune and fame. To live a most fulfilling life of uncommon prosperity, one must research and make inventions or discoveries of great value to humankind.

Researching and making invaluable inventions is an unfailing success principle to apply and pull oneself out of poverty and create great wealth for countless users. One of the best ways for wealth creation is the making of inventions via research execution that benefits all the users. The wealth that inventions bring benefits all the users of the inventions.

It is not easy to research and make inventions of great value. The making of immensely useful inventions with research is often done under unpredictable and tempestuous circumstances, as exemplified by prolific inventors.

No matter how difficult it is to research and make valuable inventions, it is much easier to execute highly demanding research, make incredible inventions and achieve true prosperity than to accept and live with poverty and misery. It is much better to research, make exceptionally useful inventions, advance the world and attain true indelible prosperity than not researching, inventing nothing, retrograding the world and being a loser forever. The world has no reason to remember and celebrate someone who invented nothing.

It is much better to sacrifice everything and invest everything in research and make valuable inventions that attract great wealth than live an unsuccessful life because one refused to research and failed to make life-changing inventions. Do the best thing. Invest in research and make life-changing inventions, discoveries, or innovations.

The best and surest way for a person to attain uncommon achievement, success and wealth is the making of valuable inventions via research. Make inventions of boundless value with your research. Then, you shall automatically achieve real success, wealth, prosperity, and fulfilment.

Be intoxicated with invention-making ambition. Crave the making of inventions far more than everything else. Then, committedly embark on research for the actualization of your cravings.

With rigorous research execution, dramatically accomplish your invention-making destiny and advance the economic well-being of the target audience and others worldwide. Let the inventions your research makes positively affect people and propel them to accelerate faster toward their full potential attainment. Let your research output transform society from what and where it is to what and where it desires to be.

ONLY THE AUDACIOUS MAKE INVENTIONS

Fire separates gold from the dross. Similarly, the difficulties and adversities encountered in the process of making inventions separate inventors from the rest of humans. Face and overcome all the adversities preventing you from astoundingly completing your invention-making journey. Get the invention your research targets made at all costs.

The power of making excellent discoveries or inventions is within everyone's reach. Every person who truly cares can research and make invaluable inventions, discoveries, or innovations for the advancement of the world. Meticulously embark on the research for inventing, discovering, or creating something of great value upon which the blissful future we crave depends.

Though everyone has within him the infinitely creative power to do research exceptionally well and make inventions, only very few people get research done skillfully and accomplish amazing inventions. Adversities separate successful researchers and inventors from all the other people.

Only the audacious people get research executed and exceptionally make inventions. A person must be extremely bold, daring, fearless, brave, and enthusiastic to execute research and get valuable inventions made profoundly.

Invaluable inventions and discoveries with research are done only by very few extremely original and daring people. Only the few who do not let prior ideas restrict them triumphantly research and make creations, inventions, discoveries, and innovations of tremendous value.

The exceedingly inventive people alone get research accomplished outstandingly. Every research work outstandingly accomplished invents something of great worth to humankind. The making of discoveries and inventions demands being intrepidly adventurous in research endeavours.

Courageously set out with your research to make a highly valuable invention. Immovably stand for making the invention which the research targets until it is marvellously achieved.

Only a person who is resiliently steadfast in the invention-making journey succeeds in making the target invention. An individual who is not firmly unyielding in his research for the making of the target invention could fall for anything and fail in achieving the goal of the research.

A man who resolutely sets out to get an invention made with his research must pay the full price for the invention-making. He committedly strives to make the invention with everything about him until it gets superbly done.

With the highest stamina level, he continues working on the

research even when he is too poor to eat. He rises very early at dawn and works on the experiment daily until sunset and long into the night after all others have retired.

The person buys all the resource materials needed for making the invention. He walks to an unimaginable length to borrow resources he cannot buy and maximally utilizes them for the research. He invests many years amid woe and wants to intensely practice and attain perfection in the craft of making inventions.

Successful invention-making is the offspring of enduring travail, toil, and drudgery. With the unconquerable will to exert every brain cell and every fibre in a person's body, he completes the research that makes an invaluable invention. It is with the courage of nonstop exertion of all the necessary actions, even in the face of death, that an invaluable invention is made.

The making of an immensely valuable invention is the hallmark of achievements. It takes the exquisite completion of every one of the various parts and sub-goals of research to make an invention. When a person rigorously completes all the several aspects of research, exceptionally attains all the sub-goals of the research and achieves the making of the target invention, he makes a name that can never die. The extremely useful inventions made by a person immortalize him on the planet.

An inventor's name never dies. The inventor of something of immense value lives forever in the minds of the people. Please research, invent, and indelibly live in the people's minds.

Making a valuable invention is the greatest service that one can render to humankind. There is no greater demonstration of a person's love for humankind than making an invaluable invention.

An individual cannot bequeath humanity any gift better than an indispensable invention. A person should invest in everything to make such inventions to advance the world and improve humankind. Unfailingly research and create, invent, discover, or innovate something that adds great value to human lives and advances the world.

DOGGED PERSEVERANCE IN INVENTION MAKING

When a person executes research with so much rigour that all the other people have not and would not, he exceptionally completes the research and makes an invention so valuable that others cannot make it. The person who exerts dogged perseverance in the thorough execution of research beyond what all others can do is the one who makes a vital invention.

The making of the invention is much more of a relentless and nonstop exertion of every needful action and investment of every treasure, talent, and time in the research than of all other factors.

Those who quit the research execution journey never make any inventions. Quitters never complete their research. Consequently, they never create a thing, discover nothing, and make no innovation.

Making valuable inventions is only for people who persistently invest all they have into the research execution until an awestruck completion. Success in invention-making belongs to only the people who increasingly invest more and much more in everything the research execution journey entails until a marvellous accomplishment.

Even geniuses, people with all the talents and research execution skills who do not tenaciously demonstrate persistence, perseverance and passionate exertion of action, cannot make inventions. They easily quit the tedious journey of invention-making.

Before a person can make an essentially valuable invention, he must exhibit the following characteristics:
1. Never ever quit.
2. An ever-increasing untiring perseverance.
3. Indomitable persistence.
4. Intensifying investment of everything in the research.
5. Escalating passionate exertion of action in the invention-making journey.

A person must persistently persevere in investing all and everything in the research execution more than has ever been done for the making of an invaluable invention before he can get it accomplished. Most often, why the particular invention has yet to be made by some other people before him is not because they never attempted making it. Rather, it is because they did not persistently persevere in investing all it takes to make the invention.

Every valuable invention made is a symbolic beacon that lightens the path for others to aspire and accelerate faster toward the attainment of their full potential. Rigorously research and make inventions of indispensable value. Let your inventions polish, spur and help others successfully embark on invention-making journeys.

EXERT THE INFINITELY CREATIVE POWER WITHIN YOU

Deeply within each person lies a hidden treasure. That hidden treasure is an infinite power for research execution and making extremely useful inventions, creations, discoveries and innovations. The deeply hidden treasure, the infinite creative power within everyone, accounts for our innate desire to solve unsolvable problems and make impossibilities possible.

From the beginning of time, humankind has kept searching, mostly in vain, for a way to ultimately express the infinite creative power latent within them. The people who have found it are the very few who thoroughly executed research and made indispensable vital inventions, discoveries, and innovations that advanced the world and improved humanity.

Research, the making of impossibility possible and unsolvable problem solvable, is the only way by which a person successfully expresses the deeply hidden treasure and infinitely creative power within him. The infinitely creative power, the deeply hidden treasure, within a person is only expressed by making inventions, discoveries, innovations or creations of inestimable usefulness through research.

Please do everything to be among the few who unleash their infinite creative power by researching meticulously and making extremely valuable inventions, discoveries, innovations, and creations that better humankind and advance the world.

We all have the infinitely creative power to make crucial inventions, discoveries and innovations to advance the world. The key to unleashing the infinitely creative power is research, meticulous research. When you thoroughly execute research, the inner treasure, the infinitely creative power within you, unlock valuable inventions that humanity and the world need to advance radically.

Set out and make the best possible use of the inner treasure, the infinitely creative power that is deeply seated within you, by meticulously researching. Then, you will be given much more of the divine spark, the infinitely creative power, to do exceedingly greater exploits in the making of far more valuable inventions.

None else can unleash your indwelling, infinitely creative power for you. No other person can tap into your boundlessly creative power and progress towards greatness for you. You alone can do the research and make the best of the inventions, discoveries, creations, and innovations your infinitely creative power is capable of.

Only through individual effort, dedication, commitment, passion and grit can you best utilize your divine spark to make uniquely valuable inventions. You can accomplish the herculean research for making the target invention only by singular focus, relentless pursuit, and ever-increasing action exertion.

Your nonstop persistent investment of time, treasure and talent alone can get the invention-making research done. No body else can do the research for the attainment of your peak potential for you.

No other individual or group can live your life for you and fulfil your life mission on your behalf. The onus of utilizing your divinely creative power in invention-making through research rests solely on you. Kindly live up to it, please.

THE GREATEST INVESTMENT TO MAKE

The greatest and best investment to make is in research. Invest a lot more in research than in all other things. Research or fund research to successfully invent something of immense worth for humankind.

Never live a life without researching or funding research to add tremendous value to human existence. A person's mission on earth is never fulfilled until he has funded others' research or personally executed research and invented something crucially vital for humanity. It is the most regrettable tragedy to leave the world without inventing something to advance the universe and improve lives and society.

Look for great human needs and invest in executing research to invent enduring practical solutions to meet some of the most important human needs. When a person neither sponsored nor executed research for making inventions that solved people's pressing problems and advanced the world, he lived a most unfortunate life.

Always put others first and add value to their lives with your research discoveries, inventions, creations, innovations, and publications. Identify great human needs. Set finding lasting solutions to them as your primary research goal. Then, fund or execute the necessary research to invent enduring solutions for the problems that have never before been solved.

Invest much of what you have in funding or carrying out research that improves people and makes them accelerate swifter to attain their full potential. Investing in research to make inventions that help others achieve what they want will consequently enable you to achieve what you want in life.

Executing research that adds great value to others' lives will automatically heighten and multiply the value of your life. When you selflessly invest in research for the invention of what assists others in advancing swiftly toward attaining their maximum potential, you successfully create a legacy that will outlive you.

With integrity and selflessness, practically answer two questions in each research endeavour.

At the onset of the investigation, ask:

"What exceptionally good thing will the invention or discovery my research aims at do to improve people, society and the world?"

On completion of the investigation, ask the evaluative question:

"What phenomenally good thing has the invention made with my research accomplished for the betterment of people, improvement of society, and advancement of the universe?"

Then, precise positive answers should be obtained to ascertain the attainment of the ultimate goal of the research.

BECOME THE LIGHT THAT ILLUMINATES OTHERS

Research and become the light that illuminates others with your discoveries, inventions and innovations. It is best to be the light that illuminates the path of others instead of merely holding a torch to light the path for others.

When you are merely holding the torch, you are not the light; you have not lit or lighted yourself. Hence, others can only get or gain a little from the torch.

On the contrary, when you are the light itself, everyone, everywhere you reach, electronically or physically, will optimally benefit from the light, which is you. Never let the light go dim at any time.

Keep shining brighter and much brighter each time. Keep producing greater effulgence from time to time.

Light yourself up with the products, solutions, inventions, discoveries, creations, and innovations you make. Only then will you find much more than success, positively affect far more people, and gain greater significance.

SEEK SYNERGY

Seek and get the benefit of synergy in some of your research endeavours. When two or more passionately dedicated persons think together and work collaboratively on achieving one research goal, it gets much better actualized and faster than the sum of solo investigations could have done it.

Each member of a collaborative research has useful ideas, experiences and the total commitment to the actualization of the goal of the investigation. The indispensable experiences that one member of the team does not have, the other member has. The resource a member does not have, another person in the collaborative research team has. The collaborative research team members complement one another for the best attainment of the research goal.

As always, two good heads that are thinking together and working in the same direction are better than one. Collaborative research execution has a relative advantage over solo research in many ways.

A wagon so heavy that one horse cannot pull is very easily driven by two or more single-minded and committed horses to its destination. It is the advantage of synergy in collaborative research that allows for faster and much better completion of the invention-making journey. A great number of inventions of extremely valuable worth are made by people in collaborative research.

Each member of a collaborative research team has to be completely committed to the achievement of the research goal. Each of the collaborative team members has to be frankly dedicated totally to the timely completion of the invention the research aims at. Every of the members has to be readily willing to make boundless sacrifices for, and unreserved investment in the completion of the invention.

The research team members must value and respect one another as oneself. The ideas and contributions of a member cannot be valued and respected more than how the other members perceive him and his worth. In other words, shared ideas and thinking in the collaborative research team can only be as good as the members of the team.

A collaborative research team of persons who have yet to individually develop themselves adequately for research execution cannot do the needful. The team can only make the desired invention or discovery when every one of its members is adequately self-developed for invention-making.

Conversely, a collaborative team of prolific or renowned

researchers is much more likely to get the research executed exquisitely and accomplish the target invention. Develop yourself fully and be part of the celebrated research team for invention-making.

CHAPTER 17

DEVIATE FROM THE STATUS QUO TO INVENT

Deviate from the status quo and established tradition to invent. The making of inventions demands the researcher to deviate from the status quo.

1. To invent, deviate from the status quo.
2. Discard popular routines to make inventions.
3. Be a positive thinker.
4. Challenge the established tradition.
5. Become a maverick inventor.

TO INVENT, DEVIATE FROM THE STATUS QUO

Change the status quo with your research. The past ways of doing things, the past routines enshrined in the status quo, are never the best ways of researching for making inventions, innovations, discoveries, and turning unsolvable into solvable and impossibilities into possibilities.

There are always better inventions, discoveries and innovations to make when we research differently to originate completely new extremely valuable things for humankind. Always think of and embark only on making inventions that have been nonexistent before your study.

Deviate from the status quo. Change from the usual way. Go in a different direction. Veer from the past ways of doing things that prevent us from inventing or creating exceptionally great solutions to certain major problems of people, society and humankind.

Diverge from the status quo to create, invent, innovate and discover with your research. A researcher has the right to disobey unjust laws, old laws, and customary rules that prevent making exceptionally good inventions. Break such old laws and routines with your research and make extremely valuable inventions to best solve the unsolvable problems of humankind.

Refusing to accept the status quo is a primary step towards invention-making. Only by changing the status quo and setting it aside can inventions be made to advance the world.

Making inventions for the radical advancement of people, society, and the world demands a break from the past, a change that repeals, eliminates or scraps the status quo. To make great, impossible, good things possible with your research, you cannot settle for the current routine ways of doing things as enshrined in the status quo.

Smash the status quo and get the immensely valuable inventions made. The researcher must shatter the status quo before he can make the crucial inventions that turn impossibilities into possibilities.

Fight an uphill battle with your research. Make worthwhile inventions that will turn impossibilities into possibilities, change the unchangeable, and solve the unsolvable. With your research, swim upstream to make inventions and discoveries that will better people, progress society, and advance the world.

Seek not the achievement of mediocre results with your research. Only execute research to attain rare and uncommon results, the type of results that advance the world radically.

Always go for the best with your research. The best can only be found beyond what the status quo permits. Never go for anything less than the very best with your research.

Remember the definition I have advanced for research and used an entire book to demonstrate. "Research is making impossibility possible." "Research is turning solvable, the unsolvable." "Research is the quest for new possibilities" (Peter James Kpolovie, 2023). The making of inventions, discoveries, creations, and innovations, which are new possibilities, do not exist in the status quo. The new possibilities can only be found beyond the confines of the status quo.

The status quo covers only things that already exist. To empirically bring solutions, products, ideas, things, devices, technologies, knowledge, or services that are currently inexistent into existence, one must think, operate, or act far outside and beyond the confines of the status quo and established tradition.

In research praxis, it is imprudence, ignorance, laziness and stupidity to do what everyone else does, has done or can do. It is only right to be original and do what others have not and cannot do. To create, invent or discover demands doing what none else has done and no one other than you can be the first to do it.

Success in research demands going outside the norm to invent, create, innovate, and discover, making possible the impossibilities and solving unsolvable problems. Embark on this journey from today with all you have got until fruition.

The past routines, past ways of doing things, and past methods, procedures and processes of executing research are never the best ways to research, making inventions, creations, discoveries and innovations. Know that there are always, much better ways of researching to make inventions, discoveries, innovations and creations of extreme usefulness.

Originate and apply new ways, strategies, methods,

procedures and processes for research execution that will far better make the target inventions, turn possible impossibilities, change the unchangeable, and make the unsolvable solvable.

The imaginably and unimaginably great things that have not been created or invented before, we can easily invent and create with our research to best meet the needs of humankind. Such research can be better when we apply new methods, unusual methods, novel procedures and advanced processes.

Everything we enjoyably use in the world is from the inventions, creations, innovations and discoveries of our fellow men and women researchers, both living and dead, who did things differently. Every invention ever made was because the inventor did things differently.

We must earnestly exert ourselves in new research execution to give the world much more in return for all we receive. We must invent and create value with our research.

Now is the time to set the status quo aside and conduct the necessary research to create values and invent solutions that will significantly better people and advance the world. Commence and get the research work going nonstop until the awesome realization of the ultimate target.

DISCARD POPULAR ROUTINES TO MAKE INVENTIONS

Discard the past ways of doing things if you are to invent something. Abandon past routines before an invention can be made. Jettison popular ideas, ways and things for you to create value. Far from popularity, getting research done and inventions made demands originality of thoughts, ideas, processes, and actions.

Popular routines add nothing new. Popular things and routines solve no unsolvable problem. Doing things the popular way cannot make an impossibility to become possible for the best good of all.

Do not value popular ideas more than great thinking. Value good, unique thinking more than popular ideas. Question popular things and ways. Going against popular ideas leads to inventions, discoveries, and creations.

Meditate, think, experiment, and do only what will best change the status quo with your research. With your research, create, invent, and discover what will engender a better world.

Abandon routine ways. Discard tradition. Shed the customary ways, for they have stagnated the world.

Think and act creatively. Think and act inventive thoughts. Be innovative. Let your research discover what no other human has found. Let the output of your research dramatically advance the world.

Develop yourself to become an unpopular thinker and freely tinker with things. Become an exceptionally creative thinker. Be regularly engaged in creative tinkering. Go the extra mile with your research to bring your novel thoughts into existence in this physical world. Let your research midwife or turn the ideal world into reality.

Following or accepting popular opinions, ideas, and thoughts prevents a person from executing creative, inventive, discovery and innovative research. The best way to prevent oneself from making inventions, discoveries and innovations with research is by doing things the way it is traditionally and popularly done by the great majority.

In research execution, it is the extreme minority that wins. Only the extreme minority that creates and makes inventions, discoveries and innovations of immense value.

The majority do not make any useful innovation, invention or discovery. The majority do not complete research. They never create anything with their research. Do not follow the majority if you are to achieve the goal of your research.

Only exceptionally strange researchers hit inventions, discoveries, innovations, and creations of extremely useful things. To succeed in research, execute it on what people never dare to do. Execute the research in an unpopular manner.

Never ever research the popular way. Do not do a thing in the popular way if making an invention is your goal. Never do the common things. Do not follow the routine way of researching; otherwise, the goal will never be achieved. Refrain from abiding by the status quo if you are to make invaluable creations, inventions, discoveries or innovations with your research.

Only execute uncommon research. Only uncommon research adds tremendous value to people's lives by successfully making the target inventions, discoveries, innovations and creations.

Use your research to create a new trend. Never pursue the existing trend with your research.

Suppose almost every person attempts research in a way that invents nothing, creates nothing, discovers nothing, and innovates nothing. It only means that what they do is wrong, not right. You must conduct research completely differently to successfully make the target invention.

The way most people try to conduct research leads to nothing because they simply follow the wrong trend. The wrong trend of research is never aimed at any value creation or invention-making.

Research execution attempts that follow the wrong trend: invent nothing, create nothing, discover nothing, innovate nothing, and create no value. Never research like the majority of people attempt to.

Execute research uniquely for value creation and invention-making. Conduct research differently to attain the target invention, creation, innovation, or discovery. Execute research uniquely to make possible impossibilities and solve unsolvable problems. Do what none else has done to create or invent things of great value.

The making of the invention with your research demands that you see and treat what most people do or can do as bad, wrong and should not be done. Becoming a distinguished researcher and inventor demands doing novel things, things that have not been done by people and which no person other than you will be the first to do. Becoming a renowned researcher and inventor demands doing things that have yet to be done by people and which you will be the first to do.

Excellence in research demands doing things differently, in a unique way that none else has done. Demonstrate your uniqueness with your research. Research and be the first to achieve a goal, an invention or a discovery that any other human has never attained.

In research, an idea is good and right only if you are the first originating it. Creating and inventing something valuable with research demands that you are the first to get it done.

An already popular idea is wrong and bad. It should never be pursued by someone who wants to become a renowned researcher or a prolific inventor. Pursuing a popular idea never leads to the making of any invention, discovery, creation, or innovation.

In research, a good goal is the one you are the first to set, pursue and achieve. Set, pursue and accomplish only good goals (goals that others have not attained or have never achieved) with your research. Then, you will become a renowned inventor over time.

If something has never been done before, that is why you should get that thing done with your research to best meet the specific needs of humankind. On the contrary, if someone has done or achieved something before, that accounts for why you should never do it again with your research. Research only on what has never been successfully done before by any human or creature.

BE A POSITIVE THINKER

Be overwhelmingly positive in thinking and act accordingly. Great achievement, successful research execution, making of immensely useful inventions, turning impossibilities into possibilities, and solving unsolvable problems are done only by positive thinkers, optimistic and precrastination researchers.

Scientific breakthroughs are never made by impossibility thinkers, procrastinators and pessimistic people. Only individuals who believe it can be done, think they can do it, and get it done make immensely valuable inventions and discoveries with their research.

Five simple lines of thought and actions are necessary for making an invention with research.
- First, think positively.
- Second, know you can do it.
- Third, be convinced you are the best and only one who should do it.
- Fourth, get it done at all costs.
- Fifth, be the first to get it done.

There is nothing too good to be invented with research. Everything good and valuable can be invented, created or discovered if we research enough. With thorough enough research execution, any good thing can and should be invented.

Anyone who truly cares can become a positive thinker. The

gem of positive thinking and corresponding action-taking is richly present inside every person. However, only the very few who deliberately develop themselves always to think positively and timely take every needful action practically demonstrate it.

Self-preparation is the basic step for successful research execution for discovery, innovation and invention making. Only the individual who best prepares himself has the will and unshakable determination to embark upon it. Invention, creating or discovering a solution to a persistent, pressing, unsolvable problem is only done by the most prepared person. Prepare yourself always to think positively and promptly take all the necessary actions.

Mentally visualize an all-important creation, invention or discovery you want to make with your research. Structure it firmly in your mind's eye and capture it explicitly on paper in both drawing and writing. Then, set all out on the research journey to practically actualize the invention.

Keep working on its actualization without giving up, even after being turned down by the experts in the field. Believe with unwaveringly committed action that you can do it; you can make the impossible possible even though no human has attempted to achieve it.

Always know that only something that has never been done constitutes an invention, creation, discovery or innovation when you get it done. Have total conviction in your vision of making the impossible and unsolvable, respectively, possible and solvable. Exquisitely do all the work, make every sacrifice, and invest everything in the invention-making project until the ultimate research goal is outstandingly accomplished.

A positive thinker creatively identifies an unsolvable, pressing, persistent problem; he is committed completely to being the first to invent a lasting, workable solution. Next, he develops a research plan or proposal on it. Finally, he thoroughly implements the research plan and accomplishes the target invention.

A research proposal is a bridge which links where and what the researcher is and has to where and what he wants to invent, create, or discover and become. The proposal increases the probability of attaining his potential success tomorrow by giving him direction and credibility today.

He thoroughly implements the invention-making Marshall Plan. He committedly works on it nonstop until the actualization of the target invention. With the invention made, the problem that was before the investigation perceived as unsolvable, an impossible problem to solve, gets solved.

Positive thinkers creatively think and do everything to materialize their ideas. Realistic thinking and committed, unstoppable action to maximize research goal attainment culminate in making the target invention, discovery, creation, or innovation. Inventions and discoveries are made with research for solving major, persistent, pressing problems and the advancement of humankind.

Explicit plan and meticulous execution which the positive thinker does, culminates in timely research goal accomplishment, invention making, lasting solution discovery and big-time making of impossibility possible.

With rigorous research alone, we empirically invent and turn the ideal we greatly yearn for into what practically or realistically exists. The ideal world we crave is turned into a present reality with thorough research.

The expectation of the ideal society and world we creatively desire can only be made a reality currently when enough intense research is done. We should get about executing the research necessary to materialize our ideal state.

Research, invention and creation of the ideal world we

earnestly desire, expect and crave can only be done with action-backed positive thinking that sets aside the status quo. We must overcome the thoughtless red tape, bureaucracy and invention-prevention regulations the status quo poses to successfully make necessary inventions and materialize the ideal society or world we crave.

By overcoming all bureaucracy, red tape and situational regulations of the status quo that limit and prevent the making of inventions, research can be done to make the ideal state which is currently impossible to become possible. We must put all the great creative ideas for invention-making to work and use everything at present to turn the ideal world into a reality with our research.

We have to invest all that is, into researching and inventing all that should be. Put in everything currently available to successfully make the desired and very much craved inventions for the best improvement of people's lives, evolution of society, and advancement of the world.

Invest all that is concrete in your research to build, create or invent the ideally desired society and world. Committedly do all the needful willingly to make the target invention with your research.

Successful making of the desired invention is certain when all the needful is done. Get everything necessary for making the invention the research targets done, and done exceptionally well.

With enough rigorous research, there is nothing so small and irrelevant that a disproportionally big and incredibly valuable thing cannot be invented from. Something big and useful can be invented from every little thing when the necessary research is done. An extremely valuable invention can be made from anything if we research it meticulously enough.

Be the person to execute the necessary research for turning a seemingly irrelevant thing into a vitally valuable product, service, invention, knowledge or idea. Yes, you can. Embark on the research now. Plus, keep working in accelerated form on it, from strength to greater strength, until a remarkable realization of the target invention.

Every waste product is a relevant raw material for the invention or making of another useful product when enough meticulous research is done for the purpose. This accounts for why over 10,000 products have been invented from crude oil or petroleum. With much more research, many more thousands could still be invented from crude oil.

Think positively and act most creatively. Smash the status quo. And invent perfect solutions to the problems that are currently considered unsolvable. Enable humankind to meet their unmet needs with the crucial products of your research.

CHALLENGE THE ESTABLISHED TRADITION

Challenge the traditional thinking and ways of doing things. No creation, invention, discovery, or innovation of extreme usefulness can be made by doing things the traditional way. If an invention could be made by doing things traditionally, people could have invented it ever since before you.

Originate completely new ways of doing things as demanded by the invention in view. Set aside the existing tradition. Do things completely differently from what the tradition requires of people.

Every invention of great worth can only be made by going against the established tradition. Refrain from following the already established tradition because adherence to the traditional ways of thinking and doing things never leads to the making of any invention, creation, discovery, or innovation.

Every inventor in human history took extreme actions contrary to the established tradition to get their inventions made. Be a leader of thought. Design the future with your anti-status quo thinking and actions.

Create new ways of thinking and doing things to get your target invention made. When your invention gets disseminated and pervades the world, the new ways of thinking and getting things done that you created shall replace the status quo and established tradition that had all along prevented the making of the invention.

Originate new and much better ways of reasoning and doing things to get your target invention, discovery, or innovation of immense value made. With your new ways of thinking, design an exceptionally superior product and take all the superhuman actions to bring the product into reality awesomely. Always challenge existing tradition and status quo to discover novel ways of accomplishing the inventions your research aims at.

The established tradition only guides people to live an average life. Living an average life never allows for the making of any invention. That is why the average people, who constitute the majority and the masses, never invent anything. They never create value for the betterment of people and the advancement of the world.

The average people think normally or averagely and take average and common actions. As a result, they never invent, discover, or create a thing. The average people have, in reality, given up on their dream, ultimate human right, duty, obligation and responsibility of making valuable inventions. They simply settle for living a normal life that does not invent anything.

Please, unlike the average people, break out of the limitations that the established tradition imposes on people. Approach life with the greatest of dreams, exert utmost actions and exceptionally fulfil your utmost human right, duty, obligation, responsibility and mission of making inventions, creations, discoveries, and innovations of great worth. Create much more value on the planet than other human minds could imagine.

Set very big goals. Enthusiastically pursue the attainment of the goals persistently until they are outstandingly accomplished. Like any good project, an invention-making journey gets nowhere without enthusiastic persistence in doing all the needful. With impetuous passion, make all the sacrifices and invest all the resources and talent to make your target invention until it succeeds excellently.

Let the great value you create be seen, acquired and predominantly used everywhere in America and throughout every country on Earth. In collaboration with some multinational corporations, mass-produce, disseminate, flood, and dominate the market everywhere with your product. Let everyone around the world know you, your product and your brand. Let your name become synonymous with science, research, technology, and invention of immense value.

Your product should be so predominantly valued and used ineliminably across the globe that it creates a new reality, a totally different reality from everything that existed before it. Let your product substantially influence and change the world positively. The impact of your invention on the world should become so overwhelming that it dwarfs even the big dream and ultimate goal you had when embarking on the research.

The breakthrough made by your invention should never make you relax or rest on your laurels. Identify a far bigger problem and passionately embark on the research to create, discover, or invent a workable solution. The solution should again break all existing world records. Without wasting any minute, exert all the distinguishing actions nonstop to devise a solution for the long concluded unsolvable problem.

Exert even more extreme actions to create a permanent solution to the unsolvable problem. Your committed actions should be such that they unfailingly create an exceptional solution to the problem.

Committedly do whatever it takes to excellently invent a perfect solution to the problem. It is by taking extreme actions, no matter what, consistently over time, that a perfect solution is invented for a problem that has traditionally been concluded to be unsolvable.

Habitually make untold sacrifices for and boundless investments in the pursuit of solution discovery for the unsolvable problem until it gets solved permanently. Exert incredible actions habitually to solve the target problem and prove the established tradition wrong for holding that the problem is unsolvable.

A better future, when you must have provided the world with a perfect solution to the currently unsolvable problem, depends upon the enthusiastic and persistent unimaginably great sacrifices, actions and investments you make today. Make all the necessary sacrifices for and investments in, and take every one of the astronomical actions to successfully create or invent a perfect solution to the unsolvable problem. Then, take possession of the blissful future you crave.

BECOME A MAVERICK INVENTOR

Become a maverick inventor with your research. Maverick inventors are highly creative and most inventive as they never respect or adhere to the conventions of their time. Most often, adherence to the conventions prevents people from making extremely important inventions. Inventions by maverick researchers radically advance the world positively.

Slavish adherence to the status quo and conventions of one's time expresses a lack of creativity, absence of constructive discontentment vision and deficiency of innovative determination. Slavish adherence to the day's convention prevents a person from being creative, inventive and innovative.

Adherence to the convention makes a person accept whatever the status quo considers impossible and unsolvable. Hence, the person does nothing to turn conventionally impossible things and unsolvable problems into possible things and solvable problems, respectively.

To create, invent, discover or innovate, a person must demonstrate constructive discontentment and creativity and go contrary to his time's status quo and conventions.

The invention, creation, innovation or discovery of great worth is only made outside the rules. Without going outside or contrary to the rules, no new thing of immense value can be created, invented, discovered or innovated.

Challenge the status quo and convention by frequently seeking novel empirical answers to questions like:
1. What if?
2. Why not different?
3. Which better way?
4. What is more for me to get done to actualize the target invention?
5. Why?
6. Why does the new technique I proposed always work?
7. How can I provide a perfect solution?
8. What should I practically do to turn possible an impossibility?
9. What new thing can I make?
10. What else can I do that has never been done before?

DEVIATE FROM THE STATUS QUO TO INVENT

To empirically answer such questions, a person must be constructively discontented with the status quo. He must be exceptionally visionary. Above all, he must embark on maverick experimentations.

The person must dedicatedly invest in research. He has got to invent or create disruptive innovations and unusual solutions with his research.

Break the traditional rules to see the yet-to-be-made inventions behind them and take them out for the best good of all those who need them. Unless the rules are broken with your research, the inventions blocked by the conventional rules cannot be made.

Creative revolutions that change the world for the better are born by rule-breaking research. Violate the rules that have prevented people from making the invention your research targets.

A person can only invent things of substantial worth after breaking the rules of the status quo with his research. Violate the rules and make the inventions that will best meet the great needs of humankind.

A person can only attain his full creative and inventive potential by breaking the rules that limit or prevent him from making the best inventions he is capable of. Never has any world record been set or broken by an individual, organization or corporation in any endeavour field without doing something differently from all the world knows about the field.

Grow rapidly to become the most inventive maverick who daringly defies the conventions and status quo to radically change the world for the best with his research.

Rules are normally broken and violated before valuable inventions are made. An individual can only set himself free from the conventional constraints of the established tradition by breaking the rules and making the inventions, discoveries and innovations blocked by conventional constraints.

We should research and remarkably increase our propensity for inventing methods of overcoming seemingly insuperable problems, challenges and obstacles.

To create, invent, discover or innovate something of extreme usefulness, one must go contrary to the customary rules or the established tradition that do not allow for making a great breakthrough. Think outside the limiting box and work beyond the confining lines to make disruptive inventions, innovations, or discoveries that cause indispensable breakthroughs in your chosen profession.

Some customary rules only serve as self-imposed limitations that prevent us from being exceedingly creative, innovative and inventive. Break such rules and make inventions, discoveries, and creations of the greatest value to humankind with your research.

Love tinkering, learning and experimenting above everything else. Tinker daily to innovate. Experiment and learn daily for invention making. Build prototypes of your creative ideas.

Be exceptionally curious deliberately. Engage in creative discontentment meditation. Read broadly. Read books that broaden your positive thinking. Read books on research methods. Read books on statistics and research data analysis. Study books on making inventions.

Make time every day for solitary, creative meditation. Then, capture the creative ideas in writing and patent drawings. Diagrammatically represent the creative idea from all angles – front, back, sideways, top, bottom, and inside.

Generate many creative opinions, ideas and tentative solutions for each persistently pressing problem. Originate several ways of exceptional innovation and improvement of each thing that could be used to solve unsolvable problems.

Experiment with countless ways of radically revolutionizing the status quo. Let the experimentations add or create much greater value than what exists.

Originate groundbreaking ideas. Also, put ideas upon ideas and rejig them to create things of much greater value to invent, innovate and discover. Work persistently on and materialize the ideas. Get the things of greater value created or invented.

Trust yourself and your creative ability. Use your instinct judiciously to generate creative thoughts, ideas and opinions for best solving each identified problem.

Fashion out new ways of doing things. Originate unimaginably good new processes for solution creation with your research.

Always see incomparably more solutions than problems. For every problem, design and originate several probable exceptional hypothetical solutions that other researchers have never tried. Next, experiment on the hypothesized novel solutions to arrive at the highly functional, effective and efficient ones in solving the problem.

CHAPTER 18

CREATION OF VALUE WITH RESEARCH

Set a globally crucial goal and meticulously pursue committedly to conclusive actualization. With complete dedication, set, pursue, and achieve a goal of creating great value for humankind with your research.
1. Create value with research.
2. Develop invention-making habit.
3. Demonstrate God's creative nature with your research.
4. Infinite creative power in every person.
5. I can, you can, every truly willing person can.
6. Inventions today make the future.

CREATE VALUE WITH RESEARCH

Let your research discover, create, invent, innovate, make possible an impossibility or solve an unsolvable problem to best meet a great need of humankind. To achieve the value creation, a person must first set a great goal and, second, persistently exert all the necessary actions with passion for its realization until it is amazingly accomplished.

Achievements are only made where there are goals clearly set. Therefore, execution of research for making an invention, discovery, innovation, or creation of anything of value first demands an unambiguous setting of that extremely useful thing as the goal.

Create value in this world. The primary purpose for the existence of each human being is value creation. Only by creating value can man fulfil his mission or purpose of existence.

Value can best be created by executing research successfully. Research is said to have been successfully executed when an invention, discovery, innovation, or creation of a valuable solution, product, service, idea, or knowledge is made. Research and make an invaluable invention or discovery that turns an unsolvable problem into a solvable one and an impossibility into a possibility.

Let every one of your research works make an invaluable invention or discovery which demonstrates that you are, indeed, an inseparable one with the creator and sustainer of the universe. Complete research that creates or innovates something of immense worth. Getting such done is the greatest service a person can render to humankind for the advancement of the world.

Be part of the power that creates and sustains the universe by executing research which invents things that add much value to the universe. The planet is best sustained with inventions and discoveries that add value to the universe and improve the quality of human existence.

Each person can become a much better individual by researching, inventing, creating, innovating, and discovering something that adds immense value to everyone else's life. Therefore, a person must advance the world by one giant leap with the outcome of his research.

Accomplish research that makes valuable inventions and discoveries to enhance humankind. Make discoveries and inventions that extend God's creation and make people better advance speedily to actualize their full potential.

Research and make inventions that will enable each person to recognize his unlimited abilities better and do the needful to accelerate towards attaining his maximum potential. Make inventions with your research that spark the right attitude in men and women to attain their maximum potential.

INFINITE CREATIVE POWER IN EVERY PERSON

Each of us has the special infinite power for researching, making inventions and furthering creation. Embark on fully utilizing the creative power within you for executing research and making invaluable discoveries and inventions that will best meet the needs of humankind and sustain, expand, and advance God's creation.

The infinite creative power within a person can only fulfil its mission if the individual committedly and passionately executes research to make possible impossibilities and turn unsolvable problems into solvable ones. With deliberate effort, a person can harness the latent, infinitely creative power within him. Please make all the needful efforts for the purpose.

Only extremely few people utilize their infinite creative power. Be one of the few who tap their creative power to become prolific inventors.

Research and make valuable inventions. Industrially mass-produce the products, solutions or services. Commercialize them and make the fortune, wealth and glory you deserve from the inventions.

Utilize your infinitely creative power to attain success. Achievement of uncommon greatness, glory, wealth and heroism is within every person's reach. It is only a question of choice: to be or not to be an inventor.

The man who maximally utilizes his infinitely creative power to research makes inventions and discoveries. He turns unsolvable problems into solvable ones and makes impossibilities possible. And he achieves uncommon wealth, greatness, glory and prosperity. Conversely, the person who does not utilize his creative power, never researches, invents nothing, discovers nothing, creates no value, and makes no indelible contribution to the advancement of the world.

No body other than you can make the decision and take the necessary actions to accomplish great achievements for you. It is your sole responsibility to take every needful action, research, make inventions and discoveries of great value and progress towards attaining real greatness, glory and wealth.

Take the needful decision now. Exert all the necessary research execution actions now. Make discoveries and inventions, and achieve exceptional greatness, glory, wealth, and prosperity.

Through individual effort in utilizing his infinitely creative power, the person attains true success. A person's consciousness of his creative power, the magnitude to which he utilizes it in researching and invention-making for the good of the masses, is what abounds him with true success.

He, who with fervent faith, relentlessly focuses on research execution and invention-making, creates great value that brings him exceptional wealth, glory and greatness that idolize and immortalize him. Research and become one of the few who attain such true success.

Examine yourself closely. Frankly answer these questions:
1. Have I fully utilized my boundlessly creative power in turning impossibility into possibility and unsolvable into solvable for the best good of humankind?
2. Do I crave becoming much more successful than I am currently?

If you have yet to optimally utilize your infinite creative power, embark on research and make inventions or discoveries of immense value. If you yearn greatly to become much more successful than you are currently, do not be content with yourself.

Execute much more rigorous research and make far greater inventions of enormous value to humanity. Plus, industrially mass-produce and commercialize the products or services. Publish and boundlessly disseminate the findings as science books.

The more your valuable research findings and products saturate the global marketplace; the more people will acquire and utilize them to solve the target problems. Then, the achievement of greater success will naturally overwhelm you.

Absolutely, every person is born with a spark of divine creativity. A person's true success on the planet depends on how much he fans that divine creative spark into reality by researching and inventing or discovering things of tremendous value for humankind.

Relentlessly fan your divine creative power to maximum realization via research, invention-making, turning unsolvable problems into solvable ones and impossible good things into possible great things for the best advancement of humanity. Doing these automatically achieves true, uncommon success, greatness, glory, wealth, and prosperity.

I CAN, YOU CAN, EVERY TRULY WILLING PERSON CAN

You and I, and indeed, everyone, have boundless potential, like the prolific researchers and inventors on the planet. Whatever good thing you wholeheartedly visualize and exert the required actions on with your research, can unfailingly get invented, discovered or created timely.

With enough meticulous research, every good thing is possible. Such research can be done directly, indirectly, or collaboratively by anyone who totally devotes himself to it. Everything of great value to people that you intensely focus on creating, inventing or discovering with your research can surely be achieved when all the necessary effort is exerted.

Therefore, always know that you can. You can achieve that invention. You can exceptionally accomplish everything of great value to humankind that you envision and execute the needful research on.

Please do as listed.
1. Have a unique creative vision of something of immense value to people, society and the world.
2. Set the materialization of that vision as the ultimate goal of your research.
3. Have the unwavering conviction that you can achieve the incredibly valuable goal and work on it with confidence.
4. Courageously embark on the research for it.
5. Execute every bit of the research with undying passion.
6. Discipline yourself to work nonstop on it till an outstanding completion.
7. Remain focused single-mindedly on doing everything that the actualization of the goal demands.
8. Develop yourself to perfection in every aspect of research execution.
9. Never stop the self-development at the point of almost mastery of any of the research execution steps. Almost successful is not success at all.
10. Exhibit pure character in the execution of every aspect of the research.

11. Complete every unit of the tasks exceptionally well, as the research demands.
12. Prioritize the execution of the sub-goals of the research and complete each task from the most crucial to the least.
13. Make every sacrifice, no matter what the research demands.
14. Value attainment of the ultimate goal of the research much more than everything else, even your life.
15. Take total responsibility for the research from the onset to its astounding completion.
16. Never procrastinate exertion of action on the research execution.
17. Committedly write originally on the investigation two to four hours every single day until completion.
18. Thoroughly execute the experiment.
19. Accurately perform the data analysis and results interpretation.
20. Successfully apply for a patent of the discovery or invention made.
21. Commercially mass-produce the product.
22. Publish the findings as a science book for easy diffusion, best preservation and use.

INVENTIONS TODAY MAKE THE FUTURE

Today is when to start inventing the things we need. Our inventions make the future for us. We can only have the future we invent today with our research.

Our inventions determine the future for us. The destiny of humankind depends only on the inventions we make today with our research.

We have the responsibility of inventing everything that will automate the ideal world we crave. Humankind can only have the type of tomorrow which our inventions today will bring. We cannot get a better tomorrow than what the inventions we make today determine.

Today we must invent all that we want tomorrow. The future we crave, we must create today with the inventions that our research accomplishes. What we invent, we shall have.

We cannot have or become more than what we invent today with our research. We must make invaluable inventions, creations, discoveries and innovations with our research today to have the blissful future we crave. Never can we attain greater value than what we create with our inventions today.

Suppose you are not already on it, commence the invention-making journey today and sustain the needful actions until accomplishment. Make today the someday you have been waiting for to create, invent, innovate or discover something of extreme value.

Embark on the invention-making journey you would execute someday today. Make today the day for the exertion of extreme actions to make the invention of your dream. Do the things that creation of great value demands you do someday, today. Let the crucial value-adding things you could do someday be done wonderfully well today.

Do marvellous work today. Accomplish great things today with your research. With your research from this day, build mighty bridges, solve unsolvable problems, and write life-changing science books. Today is the most auspicious moment to start and keep working from strength to greater strength until the awestruck completion of the essential invention, discovery or innovation.

Research today and change the course of destiny for you, people, society and the world. Today is the day to get the research done and the invention made. Today is when to start rendering that unique service which none else can provide to humanity.

Now is the only auspicious moment for the execution of the experiment for the creation of new possibilities for the best good of all. Today is the right day to do the needful to turn the unsolvable into solvable and the impossibly good things into possible great things to actualize a blissful future. Get all the work done, and done extraordinarily well today and each day henceforth.

Complete the reading of that book today. Complete the writing of your science book today. There is never a better day than today to get done the needful things. Make today count much more than all the rest you have lived.

Build the future you want with your research today. Invent your future today.

Experiment and invent everything needed to become your future self from today. Invest all you have got in making the target invention today. Do everything you have got to do this day for the remarkable completion of your value creation journey. Make the needful innovation today. Today is the best day to mass-produce, disseminate and commercialize your products and get paid for them.

Get automated payment from today for what you know. Get paid worldwide for the things you have invented, created, discovered or innovated with your research. Get payment today for the exceptional services you render to humanity. Get paid today for your research outputs.

Turn your research ideas, passions and skills into a constant income-streaming venture from this auspicious moment. Disseminate your products electronically from this day and begin the unending residual income earning.

Research and transform your skills, passion, actions, knowledge, findings, and science book publications into automated, lucrative income-streaming businesses. There is no better day than today to launch those businesses.

The very best day and time to make the inventions on which your future depends is today, and right now is the very best moment to jumpstart it. Today should never be wasted with postponement any longer. Make the necessary inventions today and live the good future life you crave from this day.

DEVELOP INVENTION-MAKING HABIT

Habit is one of the greatest determiners of invention-making. With the right habit, the habit for successful research execution, inventions, innovations, discoveries, and creations are made to make the unsolvable become solvable and turn the impossible into possible. Let your research findings inspire people with the required habits and attitude for making inventions and value creation (Kpolovie, Joe, and Okoto, 2014).

Everyone who truly wants, can develop the right habit for successful research execution that guarantees the making of inventions. Develop that habit and research exceptionally well to improve the world.

Develop the habit of love for solving seemingly unsolvable problems to best meet the needs of humankind. Cultivate the habit of constant self-development. Establish, improve and strengthen your research execution skills. Cultivate and perfect invention-making habits. Practice and habituate reading tirelessly, writing regularly, analyzing objectively, and experimenting intensely.

Grow the habit of never giving up in your research endeavours at inventing, discovering or creating solutions to people's unsolvable persistent problems until a perfect solution is found.

Habituate mastering every aspect of and skill for outstanding research execution.

Habitually read to attain expertise in the making of inventions. A simple reason is that someone who does not read and a person who cannot read are not different in their aspiration toward adding value to human existence.

Have the habit of writing committedly. There is little or no difference between those who cannot write and those who refuse to write. It is by committed frequent writing that research gets completed, published, disseminated and preserved.

How can a person habituate research execution? Just how can a person habituate research execution skills? How can a person habituate invention-making skills? How can I habituate value creation?

It is by intentionally doing something repeatedly until it becomes part of the person's routine that typically occurs subconsciously. Self-development attitude is acquired by repeatedly developing oneself for research execution and making inventions. The habit of experimenting is formed by routinely carrying out experiments until the person consistently completes one experiment after another easily.

By reading and writing ceaselessly, a person respectively develops the habits of reading and writing. Researching can only become an individual's habit when he keeps executing research successfully. By marvellously completing research routinely, over and over again repeatedly; over time, researching becomes a person's habit.

One becomes a habitual inventor by making inventions upon inventions. Similarly, a person can only be a prolific discoverer when he repeatedly makes discoveries until making discoveries becomes part of the characteristic traits that set him apart from the rest of humankind.

Successfully create one valuable thing. Then, another, and another repeated over time, and value creation becomes your habit, your way of life. Work always on innovating and significantly adding value to things. Let innovation-making be what distinguishes you from everyone else.

DEMONSTRATE GOD'S CREATIVE NATURE WITH YOUR RESEARCH

The world is said to have been orderly created by God. God created everything and made man in his own image and likeness. The creator's breath is in every person and makes everyone share in God's infinitely creative nature. It makes each person share in God's boundless wealth.

Every person is created to:
- Know all things.
- Do all things.
- Invent all things.
- Have all craved things.
- Be everywhere simultaneously.
- Achieve all set goals.
- Better all things.
- Be all-powerful.
- Further creation of all things.

With God's creative nature and substance in us, every person who truly wants can create, invent, discover, and innovate anything of great worth with his research. A person should exhibit the limitless creative power of God in him by dedicatedly researching and inventing, discovering, innovating or creating invaluable products, solutions, ideas, services, devices, and knowledge.

CREATION OF VALUE WITH RESEARCH

Be the person who utilizes his potential by answering the life-changing call to make inventions, discoveries, and innovations of great value to humankind. Live out God's infinitely creative power in you. With your research, create, invent, discover, or innovate for the best good of humankind.

Know that self-defeat, failure and quitting are no options in each research journey for creating or inventing anything of great value to humanity.

Step out courageously to embark on research and conquer every obstacle between you and the valuable product or solution the research aims to invent. Exceptionally accomplish the ultimate research goal each time.

The creator has not and can never be defeated in his quest to create whatever he desires. In like manner, you are naturally endowed infinitely to be a conqueror in every research endeavour at inventing, discovering, creating, or innovating whatever valuable product, service, solution, device, or idea you aim at.

Every individual is richly gifted at birth with unlimited potential for greatness if only he would research enough. You have the unlimited potential to research and make the impossibly good things possible and the unsolvable problems solvable. Passionately embark on the research and make that invention, discovery or innovation of immense value to humankind.

Develop yourself to the fullest. Grow your mind. Develop your research execution talents maximally. Advance the unique gifts you are endowed with. Then, live out the creative life you were born to by carrying out research to make inventions and discoveries of great value for the betterment of people's lives, society and the world.

By nature, every person craves fulfilment. The penchant to be healthy, prosperous and fulfilled characterizes every human. We seek knowledge in place of ignorance, riches in place of poverty, and friendship over loneliness. With enough rigorous research, we can achieve, in abundance, each of the things we crave.

Strive at and actually be the one who successfully executes the research for making the discoveries and inventions that will overwhelmingly meet some of the great needs of humankind. No matter what, let it be that your research works provided the perpetual solution to one of man's persistent problems.

Like prolific inventors, each of us holds within himself every potential for growth, development, maturity, expertise, competence and the know-how for impeccable research execution that will successfully invent whatever is aimed at. Let each person release that indomitable invention-making giant latent within him.

Make the decision to be the only person who will best utilize the invention-making powers in him to execute research and create, invent and discover things of great value for humankind even if all other people opt not to research. Be the one to research, make inventions and discoveries, and turn the unsolvable solvable and the impossibly great things possible for the best advancement of the world.

CHAPTER 19

GAIN THE GREATEST FREEDOM BY INVENTING

The best and only way to set oneself free from the prison of merely acquiring and using things invented by other people is to embark on research with meticulosity and make much more useful inventions for others to acquire and use. Nothing liberates a person from only acquiring others' products as his making inventions of great value for humankind. Nothing gives a person a true sense of freedom from only using other people's products than making some inventions himself for others to acquire and use.

1. Achieve peak freedom by inventing.
2. Conquer the invention-making world.
3. Who am I?
4. What it takes to invent.
5. Invest time in invention-making.

ACHIEVE PEAK FREEDOM BY INVENTING

When a person makes an invention, he feels the greatest sense of freedom and liberation from what the status quo holds. Before the person's invention, the status quo must have held that his product is impossible to make, create or invent. Making an invention gives a person untold happiness, fulfilment and a sense of being on top of the world regarding the product he invented.

It is a sense of freedom that cannot be described. It is the greatest type of freedom an individual could possibly attain. It is a type of freedom that can only be personally experienced.

Research and make a crucially useful invention that people readily acquire from all parts of the world to benefit from. Then, and only then, will you have a firsthand experience of the indescribable freedom that invention-making gives.

Embark on research and invent a product, service, technology, solution, device, idea, or knowledge that is of extreme value to humankind. Next, experience the overwhelming satisfaction, liberation and freedom that the making of an invention gives to the inventor.

On the planet, there is nothing better to crave, work extremely hard for, and achieve that is as liberating, good and satisfying as getting an invention made. Please get a product, service, idea, device, technology, or solution invented. And you will come to the full realization that life for you could have been incomplete without making that invention.

CONQUER THE INVENTION-MAKING WORLD

Develop and demonstrate moxie to conquer all the insuperable circumstances that have prevented all others from making the invention your research targets. With an unparalleled moxie, overcome all the insurmountable odds separating you from attaining the ultimate research goal. Get the invention made excellently and conquer the invention-making world.

Overcome every one of the formidable challenges that come between you and the invention in view. And exquisitely make the valuable invention your research targets. Next, you become a timelessly celebrated star worldwide in the field of the invention you made.

Devote yourself completely to an ultimate purpose – the making of a crucially valuable invention you are most passionate about. Make all the needful sacrifices for it and invest everything in it. Extremely read, write, experiment, work, and gain extensive international business experience on how to make, blossom, and timelessly sustain your invention.

These are all things you can and should enthusiastically do. Like every renowned inventor, you carry the seeds of invention-making within your psyche. Grow, apply, and explode your deeply seated invention-making seeds.

The deeply seated seeds for making inventions within you include vision, intelligence, courage, persistence, passion, grit, self-worth, and hard work. Others are creativity, purpose, focus, natural gifts, talent, skills, and personal free will for actualizing the envisioned invention, discovery, innovation, or creation.

Work tirelessly and make that invention you are most enthusiastic about. Plus, embark on and accomplish another invention of greater worth. Next, another one, and another one, until the planet gets saturated with your products.

WHO AM I?

A person is nothing other than the value he has created on the planet. What defines an individual best are the products he has invented in the world. The impossible things on Earth that he has turned to become possible things are what best say who a person is.

The perfect solutions a person has created for problems that were before him concluded to be unsolvable problems in the world are the very best answer to the question of: Who am I? A person is a function of the solutions he has invented for solving hitherto unsolvable problems on Earth.

What best defines an individual are the products he invented, created, discovered, or innovated to immaculately meet the needs of humankind. The finest and perfect definitions of an individual are the impossibilities he has made to become possibilities and the unsolvable problems he has made solvable in the world.

A person is foremostly defined by the impossibility he thinks about the most, firmly believes that he can make it possible, and committedly invests all his resources, talent, time, and treasure to practically turn it into a possibility.

An individual cannot be better known for anything than that he made a hitherto unsolvable problem become solvable.

Get an unsolvable problem of humankind solved. Create a perfect solution to an unsolvable problem. Then, you will become known and celebrated everywhere on Earth as the person who provided the solution to a particular unsolvable problem. You will best be known as the permanent solution you created for the unsolvable problem.

When you invent an excellent solution to an unsolvable problem, you will have the unique feeling of being on top of the world, the sensation of having conquered the world. The feeling of having conquered the world and being on top of the world only comes after permanently turning an unsolvable problem into a solvable one.

Create value in this world. Then, you will best be known for the value you created. A person is not anything other than the value he has created on Earth. Each individual can best be defined as the unique value he has created or invented and added to improve humankind and the world's advancement.

Please frankly check yourself in this Invention Age regarding the question:

GAIN THE GREATEST FREEDOM BY INVENTING

Who am I?
1. Without creating any value for humankind, I am a nobody.
2. I am a nobody without inventing, discovering, creating, or innovating something, a product, service, device, solution, idea, or knowledge of crucial worth for humankind.
3. Without inventing, discovering, or creating a new possibility to better meet any great need of humankind, I am a nobody.
4. I am a nobody if I refuse to fulfil my utmost human right, mission, duty, obligation, and responsibility of inventing a product of great value to humankind.
5. I am a nobody when I refuse to fund others to invent something or create value.
6. When I make no unique contribution to the betterment of human lives, improvement of society, and advancement of the world, I am a nobody.
7. I am a nobody when I have no product, solution, or service that people pay for worldwide.
8. I am a nobody when I am not by example, making people work swifter to reach their maximum potential and become their best.
9. The world can only know me for my inventions and the value I create.
10. The world cannot know me for anything good unless I have done the needful.

Now, do you want to be a somebody or a nobody on planet Earth? To be somebody, make universally crucial inventions and create globally necessary value. Now is the time to make that invention and create the greatest value for humanity.

When a person researches rigorously enough and achieves his greatest invention dream, the ultimate goal of the research, he becomes known as the unique solution or product he invented. When he becomes known globally for what he invented, he is bound to be among the happiest, most successful, most fulfilled, and most satisfied people. The person attains the peak of his self-worth and self-fulfilment. It is not a feeling that a person should miss in this life.

The zenith of prosperity a person can attain is when he invents something of indispensable value in the world. Make an extensively necessary invention. Get known everywhere by the invention you made, the solution you discovered, the product you created, and the unique service you originated. Then, attain the peak height of true prosperity.

You will become a new person now known globally as someone you were not in the past (before launching your invention). Your new personality will propel you to crave making more inventions. It will motivate you to work dedicatedly to accomplish each of the inventions you target with your research.

Making an invention of great value is the very best thing to do and achieve true prosperity. To achieve true prosperity with the making of an invention, a person must fearlessly:

1. Follow his heart.
2. Take uncommon risks.
3. Let go of the old and familiar.
4. Successfully experiment exceedingly.
5. Invest extremely in his self-development.
6. Commit everything else to the making of the target invention.
7. Work in the alarm zone to actualize the target invention.
8. Expand the expression of his true self and value.
9. Dedicatedly seek the accomplishment of broader invention goals than everything else that exists.

10. Practically focus completely on the attainment of the ultimate research goal until accomplishment.
11. Fan his unique creative spark to the fullest by making globally necessary inventions.
12. Work and achieve the invention goal at all costs.
13. Never give or accept any excuse that will prevent you from great value creation.
14. Never give up on any invention-making journey until an actualization of the goal.
15. Extremely practice the different aspects of research execution to the autopilot level.
16. Fulfil your ultimate human right, mission, obligation, duty, and responsibility of making essential inventions.

WHAT IT TAKES TO INVENT

To invent something, a person must think and do things differently from all others. Nothing can be invented by a person who typically thinks and does things the way others or most people do. How the majority do things never leads to the making of any invention.

An invention is only achieved by someone who chooses to get it done at all costs. When a person committedly exerts all the extreme actions and does all the work it entails to make an invention, it gets done.

A person achieves what he committedly seeks and invests all it takes to get it done. When a person does whatever it takes for a product to be invented, the invention-making journey gets awesomely completed.

To make an invention, a person must keep moving ahead with the exertion of extreme actions, irrespective of the obstacles that stand in his way. Only by triumphantly and victoriously fighting for the making of an invention that a person eventually get it completed exceptionally well.

Invention-making is about never quitting, no matter what until the invention gets done. Never ever quit your invention-making journey. Why others who attempted to make that invention did not succeed is solely because they quit in the process (the invention-making journey).

Get the target product invented at all costs. Make all the sacrifices for it, and invest everything else in it. Then, the product gets invented. Often, it is the last strength, the last energy you have; your only last dollar or possession could certainly be what, when invested in, will make the invention. Therefore, never hold anything back in your invention-making journey.

Learn from the undisputed champions of World Wrestling Entertainment Inc. (WWE). Every undisputed WWE champion achieved it by utilizing the last strength and the last energy magic to win each fight. The greatest lesson WWE teaches, and we should unfailingly learn, is that invention-making demands making all the sacrifices and investing everything, even the last energy, strength, resources, and time.

The making of an invention is about making a completely committed choice or decision and responsibly following it through to actualization. Execute the investigation with extreme thoroughness when you embark on research to make an invention. Get every bit of the details required done excellently without ceasing. Then, the target product gets invented at the end.

Suppose I dedicatedly decide to invent a product and work hard enough on its completion without holding back anything. In that case, it shall get done on time. If I focus completely on making the invention and invest everything in and about me into the project, the target invention gets made.

When a truly valuable invention is made, it has a special magnetism that irresistibly draws people around the globe to it. People from all continents are captivated by the invention to make the required payment for it to benefit from the product, solution, or service.

The invented product gives the inventor the feeling of being exceptionally prosperous. Everyone who truly wants can choose to invest all it takes to invent. We individually have the power over our own path to genuine prosperity by creating or inventing a thing of tremendous value for humankind.

INVEST TIME IN INVENTION-MAKING

An invention can only be made by someone who invests all his time working on delivering the target product. Extreme investment of time in the pursuit of invention-making is indispensable for an invention of great value to be made.

Everyone who truly cares has all the time required to invest in making an invention. You have all the time to accomplish the most necessary tasks for making your target invention. You have 24 hours per day, just like each of the renowned prolific inventors on the planet had to accomplish all they have invented.

Like the prolific inventors did, invest all your time in doing only the things that will best lead to the outstanding completion of your target invention until it is wonderfully achieved. Do that in each of your invention-making journeys, and you will soon become a prolific inventor.

When a person invests all his time in doing only the things that will result in the fantastic completion of his research, he soon accomplishes the target invention. Then, he embarks on another invention-making journey and accomplishes it. Next, he executes another research and achieves the target invention.

The circle continues until he attains super abundance with the making of inventions.

When you embark on research execution, invest all your quality time and everything in and about you in the project until the target invention is fabulously made. Never rest until you exceptionally fulfil your utmost right, responsibility, obligation, and duty of making the target invention of your research.

After attaining one research goal, embark on another and work with the same dedication, focus and concentration of all your time and everything in it until the target invention is impeccably actualized. Keep working like that all the way.

There is no room for resting on this planet for a person to become a renowned, prolific inventor. You could have a short time for restructuring, rededication, and recommitment of your entire time, energy, resources, talent, and everything about you to fulfil your fundamental responsibility, duty, obligation, and right of inventing and creating value.

Do whatever it takes to maximally utilize every minute of your time to make inventions, discoveries, creations, or innovations of astronomical value to humankind. Get much more done in the time you have each day for invention-making.

Like every prolific inventor, you have 60 seconds per minute, 60 minutes every hour, 24 hours daily, 7 days per week, 4 weeks in a month, and 12 months every year to invest in your invention-making journey until the dazzling completion of each target invention. You have the sole right, responsibility, obligation, and duty of creating and investing all the needful time in each of your invention-making journeys until the ultimate goal is fabulously achieved.

Do everything necessary to always fulfil the ultimate goal of each invention-making journey you embark on. Always ensure to succeed in fulfilling that onerous mission each time. Know and always operate on the principle that 'failure is never an option' in any of your research journeys.

Every moment, be engaged in doing something that incredibly completing your target invention depends upon. Ensure to invest every time in and do everything extremely well until the thunderstruck accomplishment of the invention. Work round the clock to achieve the invention that your research targets. Passionately keep getting much more done within the time you have until your goal of making an invention is marvellously achieved.

Desist from spending any time and action on anything or activity that does not contribute to your target invention's successful completion. Invest every bit of your time, energy and resources only on things and tasks upon which the monumental accomplishment of your target invention depends.

Completely utilize every bit of your time and resources for the exceptional achievement of your ultimate research goal. For the excellent attainment of the invention that is your ultimate research goal:

1. Never relax.
2. Never procrastinate.
3. Never slow down.
4. Never take it easy.
5. Never retreat.
6. Never take a break.
7. Never get distracted.
8. Never succumb to any seemingly insurmountable challenge.
9. Never fail to make significant progress daily towards the invention-making goal.
10. Never ever quit until the outstanding actualization of the invention.

Your purpose and mission, making the invention that the research aims at, should be so strong, important, compelling, consuming, and engrossing that you ceaselessly exert extreme actions toward its attainment until it is amazingly achieved. It is totally up to you and no one else to complete the making of the invention your research targets. You are solely responsible for providing a permanent solution to the problem your research is set to solve.

Achieving your invention dream or goal demands maximum utilization of your time, talent, gift, wealth, and energy for making the invention. Total commitment, dedication, and concentration of everything you have and are for the invention you have set out to make is a mandatory condition for its attainment. Fulfil your ultimate human right, mission, obligation, responsibility, and duty of making the target invention of your research. Unfailingly create crucial value for humankind with your research and advance the world.

REVIEW REQUEST
PLEASE LEAVE A REVIEW ON AMAZON.COM

You are greatly appreciated for using this book as a guide for executing research, making inventions and creating wealth.

Please take a few minutes to write an honest review on Amazon.com. Your review would be a great encouragement for the actualization of the dream of passionately motivating more people to research, invent, create value, attain prosperity, and accelerate faster to their maximum potential.

Thank you very much for your excellent effort.

Peter James Kpolovie

REFERENCES

Alamieyeseigha, D. S. P.; Kpolovie, P. J. (2013). *The making of the United States of America: Lessons for Nigeria.* Owerri: Springfield Publishing Ltd.

Ed Bernd Jr (2000). *Jose Silva's Ultramind ESP system: Think your way to success.* Career Pr Inc. Amazon.com: Jose Silva's Ultramind ESP System: Think Your Way to Success: 9781564144515: Ed Bernd Jr.: Books

Forbes (2024). *The World's Real-Time Billionaires.* Real Time Billionaires (forbes.com)

Jose Silva and Philip Miele (2022). *The Silva Mind Control Method.* Gallery Books. Amazon.com: The Silva Mind Control Method: The Revolutionary Program by the Founder of the World's Most Famous Mind Control Course: 9781982185602: Silva, José, Miele, Philip: Books

JUSTIA Patents (2024). Patents by Inventor Shunpei Yamazaki. Shunpei Yamazaki Inventions, Patents and Patent Applications - Justia Patents Search

Kpolovie, P. J. (2011). *Cognitive enhancement: Effects of lumosity training and brain-boosting food on learning.* Owerri: Springfield Publishers Ltd.

Kpolovie, P. J. (2012). Lumosity training and brain-boosting food effects on learning. *Educational Research Journals.* Vol. 2(6), 217-230. Lumosity training and brain-boosting food effects on learning (yumpu.com)

REFERENCES

Kpolovie, P. J.; Joe, A. I.; Okoto, T. (2014). Academic achievement prediction: Role of interest in learning and attitude towards school. *International Journal Humanities, Social Science and Education*; 1(11), 73-100. 10.pdf (arcjournals.org)

Kpolovie, P. J.; Lale, N. E. S. (2017). Globalization and adaptation of university curriculum with LMSs in the changing world. *European Journal of Computer Science and Information Technology*; Vol. 5(2), 28-89. Globalization-and-Adaptation-of-University-Curriculum-to-LMSS-with-the-Changing-World.pdf (eajournals.org)

Lazar, S. W. et al., (2005). Meditation experience is associated with increased cortical thickness. *NEUROREPORT*. 16(17), 1893-1897. Meditation experience is associated with increased cortical thickness - PMC (nih.gov)

Lazar, S. W.; Bush, G.; Gollub, R. L.; Fricchione, G. L.; Khalsa, G.; Benson, H. (2000). Functional brain mapping of the relaxation response and meditation. *NEUROREPORT*. 11(15), 1581-1585. Functional brain mapping of the relaxation response and meditation - PubMed (nih.gov)

Nichesss (2024). *Richest Inventors*. Richest Inventors | nichesss

Ololube, N. P.; Kpolovie, P. J.; Makewa, L. N. (2015). *Handbook of research on enhancing teacher education with advanced instructional technologies*. US: Information Science Reference. Handbook of Research on Enhancing Teacher Education with Advanced Instructional Technologies: Ololube, Nwachukwu Prince, Kpolovie, Peter James, Makewa, Lazarus Ndiku: 9781466681620: Amazon.com: Books

Peter James Kpolovie (2016). *Excellent research methods*. Patridge Publishing. Excellent Research Methods: KPOLOVIE, Peter James: 9781482824988: Amazon.com: Books

Peter James Kpolovie (2018). *Statistical approaches in excellent research methods*. Partridge Publishing. Statistical Approaches in Excellent Research Methods: 9781482878301: Business Communication Books @ Amazon.com

Peter James Kpolovie (2020). *IBM SPSS Statistics excellent guide.* Amazon KDP. IBM SPSS STATISTICS EXCELLENT GUIDE: KPOLOVIE, Peter James: 9798563947115: Amazon.com: Books

Peter James Kpolovie (2021). *Correlation, Multiple Regression and Three-way ANOVA.* Amazon KDP. Amazon.com: CORRELATION, MULTIPLE REGRESSION AND THREE-WAY ANOVA: 9798595840255: KPOLOVIE, Peter James: Books

Peter James Kpolovie (2021*). Factor Analysis: Excellent guide with SPSS.* Amazon KDP. FACTOR ANALYSIS: EXCELLENT GUIDE WITH SPSS: KPOLOVIE, Peter James: 9798705490257: Amazon.com: Books

Peter James Kpolovie (2022). *Multivariate Analysis: SPSS excellent guide.* Amazon KDP. Amazon.com: MULTIVARIATE ANALYSIS OF VARIANCE: SPSS EXCELLENT GUIDE: 9798402243668: KPOLOVIE, PETER JAMES: Books

Peter James Kpolovie (2023). *Research: Make impossibility possible.* Amazon KDP. RESEARCH: MAKE IMPOSSIBILITY POSSIBLE: KPOLOVIE, PETER JAMES: 9798364274007: Amazon.com: Books

Sade Meeks, Kar Gal and Charlotte Lillis (2023). The complete guide to omega-3-rich foods. *MEDICAL NEWS TODAY.* 15 Foods That Are Very High in Omega-3 (medicalnewstoday.com)

Sigmund Freud; James Strachey; and Peter Gay (1989). *An outline of Psycho-Analysis (The standard edition).* W. W. Norton & Company. Amazon.com: An Outline of Psycho-Analysis (The Standard Edition) (Complete Psychological Works of Sigmund Freud): 9780393001518: Freud, Sigmund, Strachey, James, Gay, Peter: Books

United States Patent and Trademark Office (2024). *Patent Public Search Basic (PPUBS Basic).* Patent Public Search Basic | USPTO

Made in the USA
Middletown, DE
25 June 2024

56294730R00170